Exact Solutions of Shortest-Path Problems Based on Mechanical Analogies

in connection with labyrinths, mazes and graph theory

GOKHAN ALTINTAS

All rights reserved. No part of this book may be reproduced or used in any manner without written permission of the copyright owner except for the use of quotations in a book review. For more information, address: gokhan.altintas@cbu.edu.tr

FIRST EDITION

Copyright © 2020 Gokhan Altintas

All rights reserved.

ISBN: 9798655831896

To my family

CONTENTS

Introduction	1
Path Finding in a Labyrinth Based on Stress Distribution of Mechanical Analogical Model	15
Path Finding in a Labyrinth Based on Displacement of Mechanical Analogical Model	29
Exact Solution of The Shortest Path in a Maze Based on Mechanical Analogies and Considerations	40
An Exact Solution Method of The Shortest Path Problems Based on Mechanical Analogy	72

Preface

In this book, approaches based on mechanical analogies are presented for the solutions of path finding problems and exact solutions of shortest path problems. Shortest path problems are of great importance not only in terms of theory but also in solutions of optimization problems in many different areas of real life. The fact that shortest path problems are spread over different areas makes it important that it is understandable, even to a certain level, by people of different branches and education levels in order to use the proposed solution methods effectively. In the preparation of this book, special attention was paid to this issue, and the familiar nature of mechanical behaviors was supported by visuals that could be easily understood by everyone, and the theory of the essence of the approach was made without allowing it to be lost due to detailed presentations of numerical methods that are already well known. The approaches presented in the book can be realized by any suitable numerical method. The numerical methods in the book are utilized in the programs commonly used in calculations and simulations of the engineering and the gaming industry. After modeling the actual optimization problems by using the presented approach, the solutions can be made with the common software running on the basis of numerical methods used in this book or with the calculation and simulation scripts coded by the researchers themselves on the same basis. Faster progress can be made in multidisciplinary working groups on the adaptation of the finite element method (FEM) based programs or rigid body dynamics (RBD) based motion engines to presented approaches. It is thought that the use of these numerical methods in their simplest forms in the problem types of interest is sufficient, which will make the use of approaches quite easy.

In this book, not even an equation was required to present topics and approaches. Because once the fiction of mechanical behaviors is designed with a natural imagination, the only thing left for the solution of the problem is the introduction of the designed model into software created on the basis of well-known numerical methods. Of course, they can also develop their own codes in relation to the subject in those who are experienced in numerical methods and programming, and this is quite easy. It would be useful to add only here, it should not be forgotten that some limitations and assumptions regarding the numerical methods to be used during the implementation of the approaches may not always allow the simulation of realistic behavior. However, the solutions to be used in such cases that may arise during the use of numerical methods are very simple and the articles that are included in the book are widely covered in situations that can be encountered in this context. In the study, the terms maze and labyrinth are frequently used. Although these two terms historically refer to some geometric forms, Graph Theory and topology also express certain definitions. It is important to understand the "labyrinth-path finding" and "maze-shortest path" relationship, especially for those who will use the methods to be presented with their engineering approach, in connection with these broadly detailed definitions in the study. For this reason, labyrinth and maze terms are frequently used in the titles and texts of the articles.

This book is organized into four chapters. The articles in each chapter are prepared independently of each other. Although the articles are independent from each other, since the approach in each chapter covers the approach in the previous chapter, reading articles in order facilitates their understanding. In Chapter 1 and 2, each path finding problem is addressed with different mechanical analogies, and there are important differences between approaches in terms of both computational cost and criteria used in the solutions. It can be used effectively in the solution of path finding problems in both approaches. Different approaches focused on the same problem type are often presented in the studies in the book, and this enables the selection of appropriate alternative approaches according to the problem types. An approach that can be applied to shortest path problems is presented in Chapter 3. Chapter 3 also provides highly detailed information and linked solutions for situations that need attention when it comes to implementing mechanical modeling and numerical methods. In Chapter 4, a very effective and simplified method based on the displacement criteria that can be used in the exact solution of the shortest path problems constructed in the light of the warnings mentioned in Chapter 3 is presented. Solutions can be reached by using RBD based motion engines in all approaches that allow access to solutions based on the displacement criteria in the articles. All approaches proposed for the shortest path problems can also be used in the solutions of path finding problems. FEM, which engineers and scientists are quite familiar with, has been widely used in presenting approaches and simulations, but RBD-based calculations also have significant advantages such as computational cost. The main reason for the predominant use of FEM as a numerical method in the examples is the fact that FEM has many parameters that allow it to be adapted to different problem types easily and is more effective in understanding the approaches. The topics in the book are quite different from my routine academic work, and the writing of the book has been a long process due to ongoing projects, studies and contributions to education. The covid19 pandemic provided the time for me to finish this book. In the forthcoming process, if my health condition permits and I have time outside of my routine workflow, I have the idea to write a book of application solutions that combine the approaches within the scope of the book with reverse engineering problems.

I hope this book will contribute to the work of researchers interested in the subject and serve as an additional toolbox that can be used in the exact solution of shortest problems.

Gokhan Altintas

Introduction

As a result of the interaction among technology, industry and economy, many problems in real life can be accounted into as a shortest path problem, which can often be expressed in economic parameters. Due to its importance and vast amount of application areas, exact solutions of the shortest path problems are of extreme importance. Although methods used for the solution of shortest path problems are relatively simple, computational costs for obtaining an exact solution, and associated necessary time for solution sometimes lower the efficacy of the methods used. In this regard, for the solution of the shortest path problems, development of several methods providing different source points as well as evaluation of their efficiencies is a matter of research which is still on track.

This book is composed of four articles involving different approaches developed for the solutions of path finding and the exact solutions of shortest path problems. Before going through the studies in the book, it would be useful to give a short introduction concerning historical evolution of subject, as well as a brief literature survey focused on several terms and notions used.

The methods developed for the solutions of path finding and the shortest path problems are based on different foundations. It is evident that, although the structure of the problems is identical or completely same, the problems as well as the researchers focused on this subject are of different origins. In this way, since past studies concentrated on this subject may be a bit disorderly and definitions belong to terminology of different disciplines, a literature survey may be confusing. Descriptions and classifications in the book are dealt mainly with the expressions of Graph theory, a branch of mathematics. In definition of the problems, terms including maze and labyrinth*, which are frequently used since the introduction of the subject to literature are also referenced.

Not only mazes provide solutions for many path finding problems

---

* In the scope of this book, maze and labyrinth terms are assigned as different meanings {"If there is just a single path to the destination the maze is called a labyrinth." A. Adamatzky, Physical Maze Solvers. All Twelve Prototypes Implement 1961 Lee Algorithm, (DOI 10.1007/978-3-319-46376-6_23), A. Adamatzky (ed.), Emergence, Complexity and Computation 24, Springer International Publishing Switzerland 2017}.

[1], their importance is due to the fact that many optimization problems can be transformed into maze forms [2, 3]*. Although their solution efforts of mazes go back to very old, this dual interaction is just one of the descriptors of the subject topicality. Solution techniques produced for mazes can be used not only for transportation, circuit design and path finding problems based on geometrical similarity, but also for the solutions of many problems derived from many different origins. As an example, in a cost analysis problem, assuming that the alternative processes for reaching the aim are the different paths of a maze, and the solution covering the branches concluding the minimum costs is the minimum cost solution of the problem.

Today the problems are handled in the scope of Graph theory, which is a branch of mathematics concentrated on definition, concept, classification and solution of current problems definable using branches and vertices. Since origin of above-mentioned problems are independent from Graph theory and based on necessities in real life, Graph theory is generally accepted to be a common language used by scientists from different origins. Within the scope of this theory, many new solution methods and improvements in existing solutions are presented with a mathematical perspective independent from the starting point of the problems under various classifications. It is believed that Graph theory first emerged from the solution proposed by Euler to the Königsberg bridges problem, which was solved in 1735 by Euler. He submitted the solution to the St. Petersburg Academy of Sciences in 1736 [1]. The graph Euler uses in his solution is very different from the position of the river, the bridge and the land. This is because Euler also benefits from topology. In this way, the technique used by Euler has become applicable to all problems in the appropriate topology. The general solution of the problem type was solved by Carl Hierholzer in 1873 [4]. Although Hierholzer's solution handles the problem as Euler's solution to Konigsberg Bridges, this study is worth reading because of the extraordinary publishing story and his interest and level of understanding in Euler's studies [3]. In the literature, it is

---

* In addition to literature including studies focused on subject, due to wider gaps among publishing dates, reevaluation of a presentation under another topic, or for a better understanding of the complex interaction among scientists involved, it is advised to go into "Graph Theory 1736-1936"[2] and "A history of discrete optimization"[3].

possible to encounter studies solving the mazes as if they are algebraic puzzles [5, 6, 7], it is hard to classify these studies in terms of current algorithmic and mathematical approaches. Perhaps, the first study that can be accepted as a solution algorithm for mazes is the study of Wiener in 1873 [8]. Successive studies including Tarry's algorithm [9] constitutes the bases of various algorithms currently used. Maze solutions include elaborate structures in literature such as those belonging to problems in literature including discrete optimization and Graph theory. Not only limited to their chronology and content, it may be a good approach to use sources which include information about the complex interaction among scientists involved in this subject, along with the historical development of solutions [2, 3]. Although terminology belonging to Graph theory is frequently used in definition of path finding and shortest path problems currently, it would be useful to present some additional definitions used in the studies made within the scope of this book. In this study, the terms labyrinth and maze which are the former definitions of graphs are used in different meanings. Solution of a maze is simply to determine path from entrance node to exit node. If there is a single solution route, maze is named a labyrinth [10]. Considering mazes instead of labyrinths, it is understood that path finding problem has multiple solutions. In this case, problem is transformed into determination of shortest path problem. In solution of shortest path problems, the intensive emergence of commonly used math based algorithmic solutions were in very narrow date range, namely from 1950 to 1960 [11-32]. The algorithmic methods including Dijkstra's [32], A* Search (A1 [33], A2 [34], A* [35]), Breadth First Search (BFS) [36, 37], and Bellman–Ford [16, 23, 25] can be classified under this scope. Due to difficulties in obtaining exact solutions of large problems in a limited period, above-mentioned methods providing satisfactorily successful or exact solutions are accepted to be generalized solutions. In addition to the aforementioned methods, many other methods such as methods arising from analogical approaches and the variants of these methods constitute a wide range of options. Due to this great variety of solution methods, it is difficult to evaluate today's literature in a way like Pollack and Wiebenson [38] or similar, even if only for shortest path problem type. In this context, the work done by Deo and Pang [39] is quite remarkable. In this study, not only the shortest path problems but also solution techniques are classified. In the study of Madkour et al. [40], they proposed a taxonomic classification for the algorithms used in the solution of shortest path problems. These and similar classifications ease the understanding of intricate literature of the subject.

There are considerably effective methods for solving the shortest path problems, outside the scope of Graph theory, and which cannot be classified directly as combinatorial or algebraic techniques. For example, it is very common to use appropriate rules and phenomena in basic science and engineering fields such as physics, chemistry, biology, fluid mechanics in solving path finding or shortest path problems with various transformations. Since the sources of the real problem and analogical simulation are different, analogical approaches greatly increase the number of methods that can be used to solve shortest path problems. Thanks to the analogical approaches, instead of dealing with the difficulties encountered in the solutions with the classical solution methods in its own domain, it is often possible to use the new techniques, which contain quite short-cuts, in the solutions. Analogical studies are used in real life, business life and many other domains. Some of the analogical approaches are in the same functional form as they are between mechanical and electrical systems (spring-resistance / damper-capacitor) and can be easily used in solutions by making simple adaptations and transformations. Dominantly based on form similarities, studies related to analogical similarities among thermal, fluid, electric, acoustic and mechanical systems are the most elaborate and consolidated problems in literature [41-44]*. Current literature in this scope includes many analogical solution methods, however nature of analogies used depend on new subdisciplines and more complex connections, rather than main equations of residual domains [45-49]. Particularly for the solution of shortest path problems, solutions proposed in the scope of mathematics and Graph theory [50-53] keep their places in current literature, and an increasing amount of interesting analogical solutions are realized [54-57]. Today, it is very difficult to define the approaches used in the solution of shortest path problems by using the expression of analogical alone. Because, the sources of simulations and algorithms used in solutions are becoming multi-centered and more complex day by day. It is possible to use the name "Unconventional computing methods" with a rather inclusive nomenclature for the methods used in the different

---

* References provided are some of the early examples in literature, however, they do not claim that they are antecedent examples.

algorithms that have emerged from the spread of multidisciplinary studies [58-63]. In this context, "Advances in Physarum Machines" [61], "Advances in Unconventional Computing I" [62], and "Advances in Unconventional Computing II" [63] books* emerged from the "Emergence, Complexity and Computation" series containing highly inspiring works. Phenomenon in study entitled "Marangoni Flow Driven Maze Solving" [64] in "Advances in Unconventional Computing II" [63] and the videos of related the phenomenon is one of the triggering factors behind the motivation to write this book. Moreover, as can be seen in studies above, majority of the solution methods to problems including, path finding and shortest path utilize definitions including labyrinth and maze in testing, problem definition and verification phases [61-63]. Modeling phenomena that are a combination of chemical, physical and other effects, such as the Marangoni effect, is even a challenge to decide on the applicability of the analogies from these models for solutions of shortest path problems. Studies in this route provide new approaches in solution of shortest path problems, constituting a wide range of a whole [65-82].

In this book, solutions of the path finding and the exact solutions of the shortest path problems are obtained by transformation into structural forms of problems, and later application of relevant mechanical analyses. Before dealing with the studies in the scope of this book, it is opted to inform the reader about solid mechanics and the reason of selection of mechanical analogies in solutions. Practical information and principles about solid mechanics are older than ancient times, namely before the invention of writing. There are many examples of structures and equipment remaining from the times in question. The reason for this achievement may come from the vast number of consistent conceptions of human being from daily experiences. Whether a person has a scientific or engineering experience or not, one can imagine that a thick rope can carry greater loads compared to that of a thin rope. Or, simple prisms can stand on their surfaces, rather than their edges or corners. These examples and many other claims that human being has interiorized many rules of solid mechanics-whether they are written or not. For this reason, it is considered that the examples provided in the scope of this book can be understood by readers which may not have a scientific background

---

* edited by Andrew Adamatzky.

on the subject.

Design, testing, analysis and simulation of mechanical systems constitute of a big ecosystem comprising many academic subbranches, either in theory or in application. In mechanics many numerical models are available for solution of problems and many mathematical methods are available for numerical analyses of models. Problems of solid mechanics can be evaluated under different topics in terms of different options and variables, which depend on their medium, physical effects applied, geometry and properties of materials. The variety in problem types also increase the number of parameters. In addition to parameters or variables used in daily mechanical analyses including force, moment, displacement, deformation and stress; other parameters belonging to different types of problems including fracture, stability and vibration are available. Additional parameters apart from mechanics-based parameters due to numerical techniques used in solutions are also available. In this way, mechanical analogies have a large number of parameters that can be adapted to the solution of many problems and focus on the characteristics of the problems.

In this study, finite element method (FEM) was used for mechanical analysis of flexible objects, and rigid body dynamics (RBD) based motion engines were used for connected rigid body simulations. FEM is an important technique in mechanical analyses either for commercial or academic means. Similarly, RBD based motion engines are widely used in game simulators, and some other mechanical system simulators. While the computer programs used in the solutions in the book are general mechanical analysis and simulation programs, the structure of the analogical approaches with related numerical methods also allow the algorithms specialized in optimization techniques to be written easily. The book includes four articles, the first two articles [83, 84] are concentrated on path finding in labyrinths by use of different mechanics-based approaches. In the first article [83], labyrinth is transformed into a structural form which the entry and exit points are defined in terms of mechanical boundaries, and the FE model of this structure is defined as an elastic structure which the original labyrinth geometry and topology is preserved. The structure is constituted of tetrahedral elements. FEM analyses of labyrinth model created a solution based on stresses on structural system. As a result of the analysis of the model using the FEM, the solution of the labyrinth was obtained based on the stresses on the structural system. In the second article [84], FEM is used in numerical solution of labyrinth, however, this time, truss elements are selected for solution. Although the FEM is used in the numerical solution of the labyrinth, but this time truss elements are selected as the element type, and differently from the approach in the first solution, the problem is modeled based on the protection of the structure's topology and large displacements allowed under load. In this approach, the parameters followed in obtaining the solution are displacements instead of stresses. Similar approach is used in the solutions of the mazes in the third [85] and fourth [86] articles, and the details of the use of this approach, which are used with some differences, are included in the related articles. In third article [85], which includes exact solutions of mazes, similar approach based on truss elements used in second article [84] is utilized. In this article, not only the solution of shortest path problems based on different types of mazes, but the effects of FEM options and different mechanical approaches on the solution is investigated in detail. Some of the approaches presented in the article are alternatives to each other, and approaches in solutions can achieve correct results with different computation costs. Assessment of different cases during application, either in constitution of mechanical analogies, or in application of numerical methods, errors prone to occur, and suggestions and evaluations oriented on prevention of these errors are widely discussed in this section. In referenced article [85], by use of RBD based simulation platform, the solution of a three-dimensional maze is also presented. Even if the numerical method used in solution of a system modeled as multiple connected rigid bodies is different, simulated behavior of system are almost same and solution set is exactly same as those in FEM solutions. The last article [86] in book is based on second [84] and third [85], taking the issues worthy of notice in third article [85] into consideration, and presents a robust method similarly applicable to almost all types of problems. With this approach, the process of obtaining exact solutions of shortest path problems belonging to different types of geometrical and topological forms has become extremely easy. Although articles in book can be individually evaluated, in order to better understand the importance and necessity for the method in fourth article [86], it is advised to read the articles in sequence.

The solution techniques presented in the book are not only suitable for algorithm writing, but also have the feature to be easily implemented with existing structural analysis programs. Except for the approach used in the first article, it is also possible to use simulations of RBD based motion engines in approaches to solutions obtained using the FEM in three other articles. Although RBD based methods provide relatively simple and fast solutions, FEM includes a variety of parameters stemming from mechanics and numerical solution technique, which can comfortably be used in solution of more complex problems. RBD based simulation software used currently is generally in game engines. It should not be discarded that the software is generally oriented to for visual aims, rather than numerical precision. On the other hand, FEM serves for the basis for numerical evaluations of design and analysis of structural mechanics problems. In this regard, programs for computer based numerical analyses of large problems in structural mechanics generally utilize FEM.

The approaches presented in this book provide the exact solutions based on mechanical analogy. Particularly, these approaches providing exact solutions in a reasonable calculation period are not tools solely for oriented to academia, it is considered that many problems in real life can be solved using these approaches due to their ease of application and adaptability.

References

[1] Euler L (1736) Solutio problematis ad geometriam situs pertinentis. Commentarii academiae scientiarum Petropolitanae 128–140

[2] Biggs NL, Lloyd EK, Wilson RJ (1976) Graph Theory: 1736–1936. Clarendon Press, Oxford

[3] Durnová H (2004) A history of discrete optimalization. Mathematics throughout the ages II 51–184

[4] Hierholzer C, Wiener C (1873) Über die Möglichkeit, einen Linienzug ohne Wiederholung und ohne Unterbrechung zu umfahren. Mathematische Annalen 6:30–32

[5] Lucas É (1882) Récréations mathématiques. 4 vols. Gauthier-Villars, Paris

[6] Ball WWR (1892) Mathematical recreations and problems of past and present times. Macmillan and Company

[7] Ahrens W (1901) Mathematische unterhaltungen und spiele. B.G. Teubner, Leipzig

[8] Wiener C (1873) Über eine Aufgabe aus der Geometria situs. Mathematische Annalen 6:29–30

[9] Tarry G (1895) Le problème des labyrinthes. Nouvelles annales de mathématiques : journal des candidats aux écoles polytechnique et normale 14:187–190

[10] Adamatzky A (2017) Physical maze solvers. All twelve prototypes implement 1961 Lee algorithm. In: Emergent computation. Springer, pp 489–504

[11] Luce RD (1950) Connectivity and generalized cliques in sociometric group structure. Psychometrika 15:169–190

[12] Shimbel A (1951) Applications of matrix algebra to communication nets. The bulletin of mathematical biophysics 13:165–178

[13] Lunts AG (1952) Algebraic methods of analysis and synthesis of relay contact networks. Izvestiya Rossiiskoi Akademii Nauk Seriya Matematicheskaya 16:405–426

[14] Trueblood DL (1952) Effect of travel time and distance on freeway usage. Public Roads 26:241-250

[15] Shimbel A (1953) Structural parameters of communication networks. The bulletin of mathematical biophysics 15:501–507

[16] Shimbel A (1954) Structure in communication nets. In: Proceedings of the symposium on information networks. Polytechnic Institute of Brooklyn, pp 119–203

[17] Orden A (1956) The Transhipment Problem. Management Science 2:276–285

[18] Robacker JT (1956) Min-max theorems on shortest chains and disjunct cuts of a network. Rand Corporation

[19] Rosenfeld L (1956) Unusual problems and their solutions by digital computer techniques. In: Papers presented at the February 7-9, 1956, joint ACM-AIEE-IRE western computer conference. pp 79–82

[20] Ford Jr LR (1956) Network flow theory. Rand Corp Santa Monica CA

[21] Leyzorek M, Gray RS, Johnson AA, et al (1957) Investigation of Model Techniques–First Annual Report–6 June 1956–1 July 1957–A Study of Model Techniques for Communication Systems. Case Institute of Technology, Cleveland, Ohio

[22] Minty GJ (1957) Letter to the editor—A comment on the shortest-route problem. Operations Research 5:724–724

[23] Moore EF (1959) The shortest path through a maze. In: Proc. Int. Symp. Switching Theory, 1959. pp 285–292

[24] Dantzig GB (1957) Discrete-variable extremum problems.

Operations research 5:266–288
[25] Bellman R (1958) On a routing problem. Quarterly of applied mathematics 16:87–90
[26] Bock F, Cameron S (1958) Allocation of network traffic demand by instant determination of optimum paths. Operations Research 6:633–634
[27] Dantzig GB (1960) On the shortest route through a network. Management Science 6:187–190
[28] Gallai T (1958) Maximum-minimum sätze über graphen. Acta Mathematica Hungarica 9:395–434
[29] Hoffman W, Pavley R (1958) Applications of digital computers to problems in the study of vehicular traffic. In: Proceedings of the May 6-8, 1958, western joint computer conference: contrasts in computers. pp 159–161
[30] Minty GJ (1958) Letter to the Editor—A Variant on the Shortest-Route Problem. Operations Research 6:882–883
[31] Berge C (1958) Théorie des graphes et ses applications Dunod. Paris
[32] Dijkstra EW (1959) A note on two problems in connexion with graphs. Numerische mathematik 1:269–271
[33] Nilsson N (1965) Some growth and ramification properties of certain integrals on algebraic manifolds. Arkiv för matematik 5:463–476
[34] Green CC, Raphael B (1967) Research on intelligent question-answering system. Stanford Research Inst Menlo Park CA
[35] Hart PE, Nilsson NJ, Raphael B (1968) A formal basis for the heuristic determination of minimum cost paths. IEEE transactions on Systems Science and Cybernetics 4:100–107
[36] Zuse K (1972) Der Plankalkül. Gesellschaft für Mathematik und Datenverarbeitung 96–105
[37] Lee CY (1961) An algorithm for path connections and its applications. IRE transactions on electronic computers 346–365
[38] Pollack M, Wiebenson W (1960) Solutions of the shortest-route problem—a review. Operations Research 8:224–230
[39] Deo N, Pang C-Y (1984) Shortest-path algorithms: Taxonomy and annotation. Networks 14:275–323
[40] Madkour A, Aref WG, Rehman FU, et al (2017) A survey of shortest-path algorithms. arXiv preprint arXiv:170502044
[41] Hecht H (1939) Schaltschemata und Differentialgleichungen

elektrischer und mechanischer Schwingungsgebilde. Leipzig

[42] Gehlshöj B (1947) Electromechanical and electroacoustical analogies, Thesis Copenhagen

[43] Schönfeld JC (1954) Analogy of hydraulic, mechanical, acoustic and electric systems. Applied Scientific Research, Section A 3:417–450

[44] Bosworth RCL (1948) Thermal mutual inductance. Nature 161:166–167

[45] Vidović D, Sutlović E, Majstrović M (2019) Steady state analysis and modeling of the gas compressor station using the electrical analogy. Energy 166:307–317

[46] Ismail N, Ghaddar N, Ghali K (2018) Electric circuit analogy of heat losses of clothed walking human body in windy environment. International Journal of Thermal Sciences 127:105–116

[47] Chochowski A, Obstawski P (2017) The use of thermal-electric analogy in solar collector thermal state analysis. Renewable and Sustainable Energy Reviews 68:397–409

[48] Goudarzi B, Mohammadmoradi P, Kantzas A (2018) Direct pore-level examination of hydraulic-electric analogy in unconsolidated porous media. Journal of Petroleum Science and Engineering 165:811–820

[49] Endo Y, Ngan CL, Nandiyanto AB, et al (2009) Analysis of fluid permeation through a particle-packed layer using an electric resistance network as an analogy. Powder technology 191:39–46

[50] Bonsma P (2017) Rerouting shortest paths in planar graphs. Discrete Applied Mathematics 231:95–112

[51] Mozes S, Nussbaum Y, Weimann O (2018) Faster shortest paths in dense distance graphs, with applications. Theoretical Computer Science 711:11–35

[52] Cabello S, Jejčič M (2015) Shortest paths in intersection graphs of unit disks. Computational Geometry 48:360–367

[53] Asplund J, Edoh K, Haas R, et al (2018) Reconfiguration graphs of shortest paths. Discrete Mathematics 341:2938–2948

[54] Zhang X, Zhang Y, Hu Y, et al (2013) An adaptive amoeba algorithm for constrained shortest paths. Expert Systems with Applications 40:7607–7616

[55] de Almeida JPLS, Nakashima RT, Neves-Jr F, de Arruda LVR (2019) Bio-inspired on-line path planner for cooperative exploration of unknown environment by a Multi-Robot System. Robotics and Autonomous Systems 112:32–48

[56] Yang H, Wan Q, Deng Y (2019) A bio-inspired optimal network division method. Physica A: Statistical Mechanics and its Applications 527:121259

[57] Dhiman G, Kaur A (2019) STOA: A bio-inspired based optimization algorithm for industrial engineering problems. Engineering Applications of Artificial Intelligence 82:148–174

[58] Svozil K (1998) First International Conference on Unconventional Models of Computation UMC'98. An Unconventional Review. arXiv preprint quant-ph/9801060

[59] Adamatzky A International Journal of Unconventional Computing, ISSN:1548-7199

[60] McQuillan I, Seki S (eds) (2019) Unconventional Computation and Natural Computation: 18th International Conference, UCNC 2019, Tokyo, Japan, June 3–7, 2019, Proceedings. Springer International Publishing, Cham

[61] Adamatzky A (ed) (2016) Advances in Physarum Machines. Springer International Publishing, Cham

[62] Adamatzky A (ed) (2017) Advances in Unconventional Computing-I. Springer International Publishing, Cham

[63] Adamatzky A (ed) (2017) Advances in Unconventional Computing-II. Springer International Publishing, Cham

[64] Suzuno K, Ueyama D, Branicki M, et al (2017) Marangoni flow driven maze solving. In: Advances in Unconventional Computing-II. Springer, pp 237–243

[65] Dorigo M, Stutzle T (2004) Ant Colony Optimization. Bradford Company, MA. USA

[66] Nakagaki T, Yamada H, Tóth Á (2000) Maze-solving by an amoeboid organism. Nature 407:470–470. https://doi.org/10.1038/35035159

[67] Siriwardana J, Halgamuge SK (2012) Fast shortest path optimization inspired by shuttle streaming of physarum polycephalum. In: 2012 IEEE congress on evolutionary computation. IEEE, pp 1–8

[68] Tero A, Kobayashi R, Nakagaki T (2007) A mathematical model for adaptive transport network in path finding by true slime mold. Journal of theoretical biology 244:553–564

[69] Zhang X, Zhang Y, Hu Y, et al (2013) An adaptive amoeba algorithm for constrained shortest paths. Expert Systems with Applications 40:7607–7616

[70] Nakagaki T, Yamada H, Toth A (2001) Path finding by tube

morphogenesis in an amoeboid organism. Biophysical chemistry 92:47–52

[71] Nakagaki T, Iima M, Ueda T, et al (2007) Minimum-risk path finding by an adaptive amoebal network. Physical review letters 99:068104

[72] Zhang X, Zhang Y, Zhang Z, et al (2014) Rapid Physarum Algorithm for shortest path problem. Applied Soft Computing 23:19–26

[73] Zhang X, Adamatzky A, Yang X-S, et al (2014) A Physarum-Inspired Approach to Optimal Supply Chain Network Design at Minimum Total Cost with Demand Satisfaction. arXiv preprint arXiv:14035345

[74] Steinbock O, Tóth Á, Showalter K (1995) Navigating complex labyrinths: optimal paths from chemical waves. Science 267:868–871

[75] Steinbock O, Kettunen P, Showalter K (1996) Chemical wave logic gates. The Journal of Physical Chemistry 100:18970–18975

[76] Adamatzky A (2009) Hot ice computer. Physics Letters A 374:264–271

[77] Lagzi I, Soh S, Wesson PJ, et al (2010) Maze solving by chemotactic droplets. Journal of the American Chemical Society 132:1198–1199

[78] Suzuno K, Ueyama D, Branicki M, et al (2014) Maze solving using fatty acid chemistry. Langmuir 30:9251–9255

[79] Cejkova J, Novak M, Stepanek F, Hanczyc MM (2014) Dynamics of chemotactic droplets in salt concentration gradients. Langmuir 30:11937–11944

[80] Reyes DR, Ghanem MM, Whitesides GM, Manz A (2002) Glow discharge in microfluidic chips for visible analog computing. Lab on a Chip 2:113–116

[81] Dubinov AE, Maksimov AN, Mironenko MS, et al (2014) Glow discharge based device for solving mazes. Physics of Plasmas 21:093503

[82] Pershin YV, Di Ventra M (2011) Solving mazes with memristors: A massively parallel approach. Physical Review E 84:046703

[83] Altintas G (2020) Path Finding in a Labyrinth Based on Stress Distribution of Mechanical Analogical Model. In: Exact Solutions of Shortest-Path Problems Based on Mechanical Analogies. Amazon, pp 15-28

[84] Altintas G (2020) Path Finding in a Labyrinth Based on

Displacement of Mechanical Analogical Model. In: Exact Solutions of Shortest-Path Problems Based on Mechanical Analogies. Amazon, pp 29-39

[85] Altintas G (2020) Exact Solution of The Shortest Path in a Maze Based on Mechanical Analogies and Considerations. In: Exact Solutions of Shortest-Path Problems Based on Mechanical Analogies. Amazon, pp 40-71

[86] Altintas G (2020) An Exact Solution Method of The Shortest Path Problems Based on Mechanical Analogy. In: Exact Solutions of Shortest-Path Problems Based on Mechanical Analogies. Amazon, pp 72-85

Path Finding in A Labyrinth Based on Stress Distribution of Mechanical Analogical Model

Abstract

Path finding solution of a maze is to determine the way from input node to output node. If there is a no multiple edges between any two nodes, maze is named a labyrinth. Labyrinth solutions are more than just the determination of the path between the entry and exit points of a geometric form. Because the techniques used in path finding solution of labyrinths have the potential to be applied to many real-life problems.

In this study, an approach based on a mechanical analogy which can be used in solution of labyrinth problems is proposed, and the numerical solution constituted in this approach is based on finite element method (FEM). The labyrinth is transformed to a structural model consisting of a FEs, and tetrahedral elements were used since they have the potential to protect geometry under mechanical boundary conditions and topology of the structure, whereas the original model geometry can be transferred to the model precisely. Static analysis procedure was used in finite element analysis (FEA) of problem and path of labyrinth was determined based on stress distributions in system under certain mechanical boundary conditions.

The stress distribution in structural system was effectively obtained by reducing the solution to a single stiffness matrix by using FEM. In the study, four solutions for different entry and exit points on the same labyrinth system are presented. The relationship between stress distributions resulting from FEA and the solution paths has been tried to be presented with visuals that even those who are not familiar with mechanical analysis can understand.

Additionally, the presented approaching has many tunable parameters that can be adapted depending on the different problem types, and it is thought that the problems in the context can be handled with a different perspective by using this approach.

1 Introduction

Labyrinth is a simple form of maze, having a single solution path. The simplest definition of path finding in the labyrinth is to determine the path between enter and exit nodes. Today, these two terms which are topologically distinct, can often be used in the same meaning in literature and historical records [1]. Since labyrinths are special forms of mazes, solution methods available for mazes can also generally be used for solutions of labyrinths.

Labyrinths and mazes have attracted the attention of humanity in terms of religion, aesthetics and other aspects throughout history, and many works that have been made in the past contain these structural geometric forms [2, 3]. However, currently, in addition to problems stemming from geometrical nature of the problem (e.g. transport, computer games, electronics and robotics), many important problems including work flow plans and cost optimization can also be evolved into maze and labyrinth forms.

Although the resources regarding solutions to labyrinth problems are different [4-12], in evaluation of techniques, the time necessary for solution and computational cost criteria are initially considered. Although the origins of many of the methods commonly used in solutions today are long before the common use of Graph theory [13], they are now predominantly expressed by definitions of Graph theory [14-20].

Development of these methods and their varieties, which provide solutions within a reasonable period and precision, are still going on. Apart from direct mathematical and combinatorial approaches, techniques based on analogies utilizing rules and processes belonging to biological, chemical, physical and methods not directly from pure mathematical fields provide novel approaches for solution of labyrinth and maze problems [21, 22, 23].

In this study, the solution of labyrinths is obtained by use of an approach based on mechanical analogy. In the approach used, not only the topology of the labyrinth, but also the geometry was preserved, precise solutions were obtained, and the numerical solution of the problem was reduced to the solution of a single matrix obtained by the application of FEM.

2 Method and preliminary solution

Before starting the solution of the main problem, it would be useful to give a brief preliminary information with a simpler geometry instead of the problem geometry under consideration, in order to make the approach to be used more understandable. Although it has a very simple geometry, Y-mazes are used in various types of problems, such as observing animal learning and decision-making behaviors, or examining the path finding behavior of plant roots [24], [25], [26]. In this context, Y-maze has been chosen for preliminary presentation of mechanical analogy used.

The stress analysis of this system can be easily solved by using the assumptions and equations of strength of materials course without the need for any computer procedure. The procedure in the simple manual solutions is to obtain internal forces from external forces and to obtain stresses from internal forces.

If the system's geometry and boundary conditions are complex, it may not be easy to reach a solution using analytical methods. In such a case, FEM is used predominantly in solutions. FEM as numerical calculation technique is used in numerous fields of engineering and science. The technique provides efficient and precise modelling for the behavior of thermal, mechanical, fluid or other complex phenomena. The method discretizes a complex domain into assembly of nodes and elements of simpler shape and equations, called a FE mesh. Due to the fact that the FEM is so common, there are many commercial, or freeware programs that can be used in solutions [27]. Researchers, engineers or people interested can also write their own codes and programs.

In Fig. 1, as the preliminary solution, there is an encastered support in the upper and a load in the lower left part of the Y-maze. The solution of the problem was made by static analysis procedure, where the equation set of the stiffness matrix was easily solved in single step and 32532 linear hexahedral elements were used in the discrete model. Y-maze geometry fits into the outer rectangular prism with a size of 5x6x0.8 cm^3. The selected material in analysis is steel having a modulus of elasticity of 195 GPa. The distributed load with a total value of 0.1 N was applied as shown in Fig. 1.

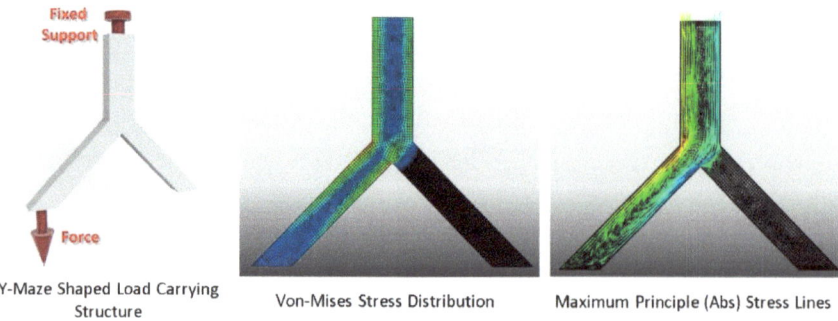

Fig. 1 Y-Maze structure and stress distribution after loading

Stress distributions as a result of loading and support conditions of the simple system in the form of Y-maze are presented in Fig. 1. According to results, it is clear that the stress distribution of the Y-maze occurs between the support and the loading locations. The reason for this form of stress distribution is due to the topological properties of geometry.

The Y-maze is a labyrinth-type maze with a zero-genus value and therefore a single solution is possible. As long as the small deformation theory and linear behavior is valid and the topology is preserved, this behavior does not change, no matter how complex the labyrinth is.

As with this solution, for all values of the load and material parameters where the small deformation theory is valid, the values of stresses and deformations may vary but the distribution form is stable or nearly unchanged.

Consequently, no matter how complex the labyrinth geometry is, the stress distribution will always be on the path between the support and the load zones.

3 Definition of labyrinth problem

The labyrinth geometry with points A and B is presented in Fig. 2. As a solution to the problem, it is necessary to determine the path between these two points. It should be known that if any physical link exists between any two points of labyrinth as shown in Fig. 2, it is always possible to make a solution by means of mechanical approach presented in the scope of this study.

In order to solve the problem, awareness or prediction about the Graph Theory parameters or topology parameters of geometry is not necessary. However, classifications based on these parameters are useful in the selection of mechanical analogies to be used in modeling of problems and the numerical methods to be used in solution.

Fig. 2 Labyrinth geometry and the points to be connected with the route to be determined

As the geometry in Fig. 2 is relatively small, some features are immediately noticeable. At first glance, one of the properties noticed is that there are no multiple paths between points A and B. In the topological perspective, the genus of the geometry is zero. If the entire geometry was made of a soft material, the shape could be converted into a sphere with points A and B on its surface.

This information shows that there is only one route connecting the points A and B, in other words, the maze can be called labyrinth, as in the preliminary problem.

4 Analysis and results

Although numerical values do not have a decisive effect on determining the path of stresses, the material and geometry properties should be chosen as realistic as possible for the numerical stability of the FEM solution.

For this purpose, in order to constitute a relatively solid structure which will show small deformation under a small amount of external loading, the scale of the labyrinth geometry was changed to fit the volume of the outer rectangular prism with a size of 8x3.6x0.2 cm^3.

Selected material for simulation is a steel with 190 GPa Young's modulus. The distributed load with a total value of 0.5 N was applied at the region and direction shown in Fig. 3.

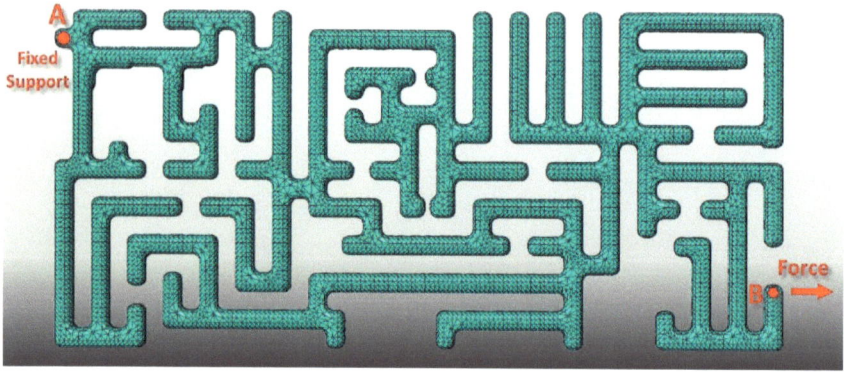

Fig. 3 Finite element mesh of the labyrinth

The analysis in this section was achieved by using a FE model with a large number of elements (113790 quadratic tetrahedral elements) to make it visually more understandable.

Definitely, this problem can be resolved by utilizing a model with lower number of elements, in comparison with that in current solution.

In Fig. 4, the boundary conditions of the main problem and the results of the mechanical analysis are presented. The solution path of the labyrinth connecting the two points specified in the boundary conditions was easily determined based on the stress distribution.

Due to the fact that there are no multiple edges between points A and B, this stress path is the only solution of the labyrinth. In this calculation, the main numerical solution was obtained with the direct solution of a single matrix by use of the static analysis procedure.

Load and support locations

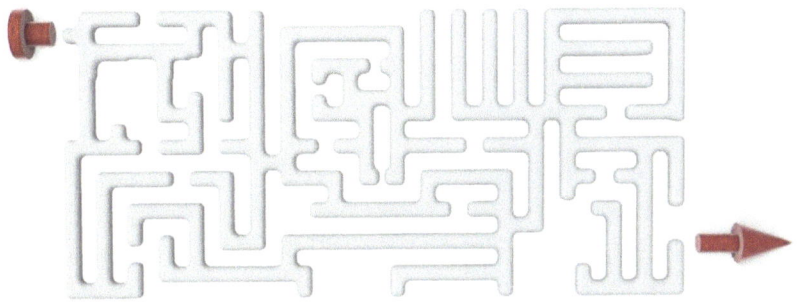

Stress path between load and support locations (Von-Mises stress distribution)

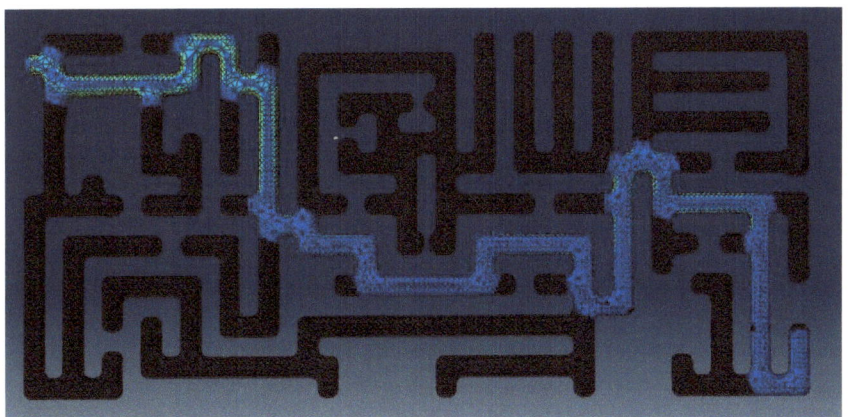

Fig. 4 Model based on mechanical analogy and solution of main labyrinth problem

For the same labyrinth geometry, reinforcement analyses were performed for different support and loading locations. The results are presented in the Fig. 5, Fig. 6, and Fig. 7. The result of the Reinforcement Solution-I in Fig. 5, which is intended to determine the path between two very close points, is perhaps the result of the stress distribution behavior which is most understandable.

Load and support locations

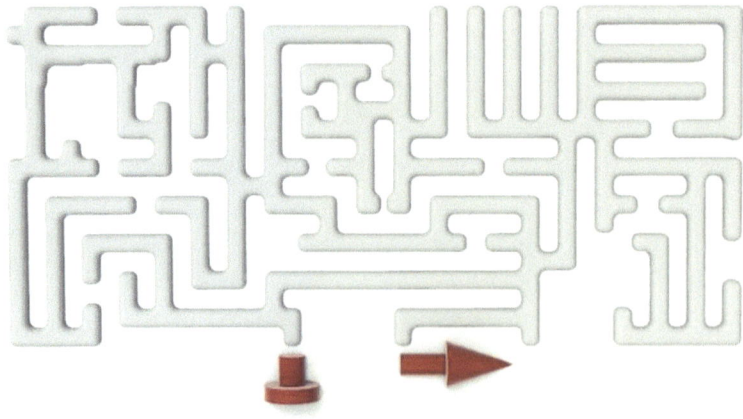

Stress path between load and support locations (Von-Mises stress distribution)

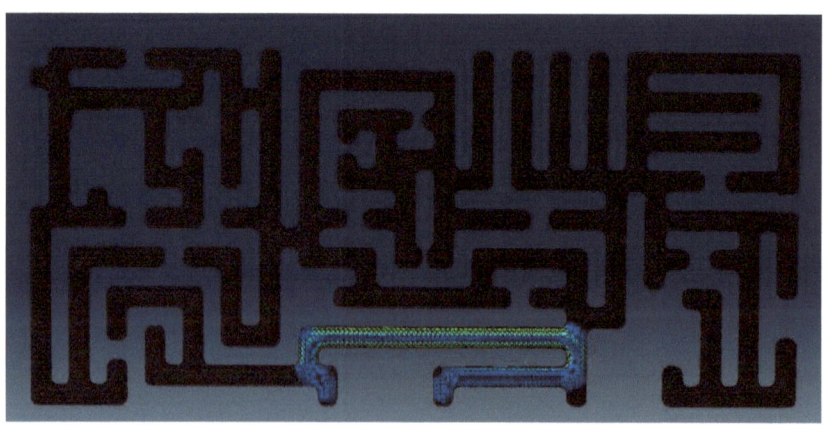

Fig. 5 Reinforcement solution-i of labyrinth problem

Load and support locations

Stress path between load and support locations (Von-Mises stress distribution)

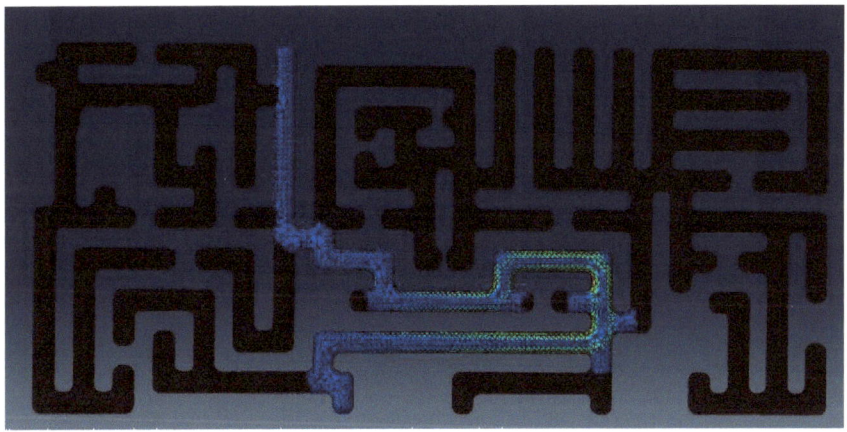

Fig. 6 Reinforcement solution-ii of labyrinth problem

Load and support locations

Stress path between load and support locations (Von-Mises stress distribution)

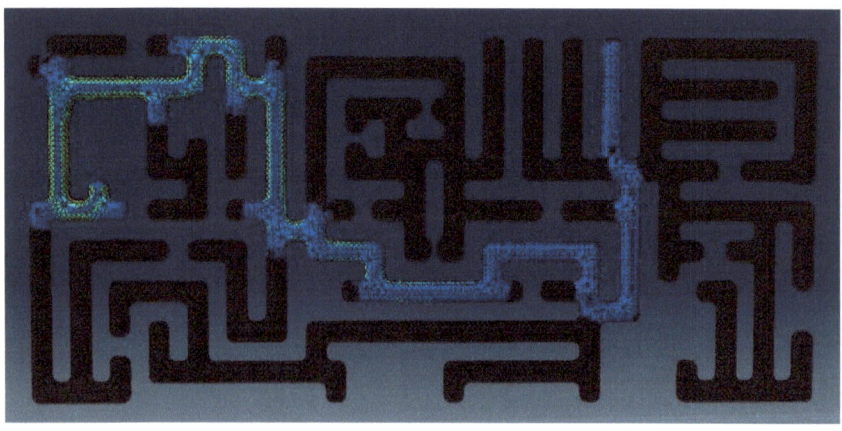

Fig. 7 Reinforcement solution-iii of labyrinth problem

As can be seen from the solutions presented, since the systems are in the form of a labyrinth, the length and thickness of the branches expected to affect the stress distribution in other geometries do not cause any significant stress distribution in the regions outside the solution path. Although this situation verifies the accuracy of the theoretical approach, during application of dynamics analyses instead of static, inertial effects may cause additional stresses and deformations which may also bring complications in convergence to a solution.

5 Conclusions

In this study, an approach based on a mechanical analogy that can be used in solutions of path-finding problems of labyrinths is presented. This approach is based on the mechanical analysis of the resulting system by converting the geometry of the labyrinth into a structural system capable of load bearing and the entrance and exit regions of the labyrinth to suitable boundary conditions. For obtaining a fast and easily converging solution, the structural properties and boundary conditions of structural model are designed in accordance with the static analysis procedure which depend on theoretical assumptions based on validity of the small displacement and deformations. The approach in this study completes the numerical analysis procedure it is aimed to obtain the solution as fast as possible. FEM and tetrahedral elements were used in analyses. The type of element used to be able to fully reflect the mechanical model to the analysis is vital, and it is important that all parts of the labyrinth in its original volumetric form are composed of elements whose surface contact is maintained to a degree that does not show the hinge feature. In other words, superficial continuity zones in real form of labyrinth should not solely be reduced to edge or point continuities. Above mentioned undesired cases can result in dynamic mechanisms which do not exist in original form of labyrinth, which also cause misutilization of static analyses procedures far from the way they should be used.

In this study, the solution was determined based on the stress distribution that occurred in structural form as a result of mechanical analysis. It can be seen from the results that no significant stresses arise in the other parts except the solution path connecting the entrance and exit points of the labyrinths. Von-Mises stress distributions were used in the visualization of stress distributions by focusing on the scalar values of stresses as much as possible.

Even in the most basic FE analyzes, many parameters such as deformations and displacements are included in the results obtained as well as stresses. This study could also utilize deformations instead of stresses for solutions. However, excluding special cases, obtaining a solution of labyrinth using displacements by strictly adhering to the current method is very difficult to use. It is important that the parameters of the mechanical model, in which the original problem is transformed with the preservation of the geometric and topological properties, remain at the static solution limits of the problem and, in connection with this, be carefully selected for the stability of the matrices. In this scope, although the static analysis procedures are to be applied assuming that the displacements are relatively small, it is not always possible to obtain this behavior in greater or complex problems, therefore, FEM analyses should be employed free of interaction aspects.

In this study, Visual intelligibility is prioritized and therefore the models used in solutions are not optimized in terms of computational costs. The approach in this study can be employed by use of commercial and freeware FEM software, or it can be specifically coded and applied to any path finding problem within the scope of the subject.

References

[1] Adamatzky A (2017) Physical Maze Solvers. All Twelve Prototypes Implement 1961 Lee Algorithm. In: Adamatzky A (ed) Emergent Computation. Springer International Publishing, Cham, pp 489–504. https://doi.org/10.1007/978-3-319-46376-6_23
[2] Matthews WH (1922) Mazes and Labyrinths. The Ballantyne Press Spottiswoode, Ballantyne & Co. Ltd., Colchester, London & Eton
[3] Doob PR (1992) The Idea of the Labyrinth: From Classical Antiquity Through the Middle Ages, 1st edition. Cornell University Press, Ithaca u.a
[4] Levitt G, Rosenberg H (1987) Differentiability and topology of labyrinths in the disc and annulus. Topology 26:173–186. https://doi.org/10.1016/0040-9383(87)90057-7
[5] Al-Muhammad J, Tomas S, Ait-Mouheb N, et al (2019) Experimental and numerical characterization of the vortex zones along a labyrinth milli-channel used in drip irrigation. International Journal

of Heat and Fluid Flow 80:108500. https://doi.org/10.1016/j.ijheatfluidflow.2019.108500

[6] Goldstein RE, Muraki DJ, Petrich DM (1996) Interface proliferation and the growth of labyrinths in a reaction-diffusion system. Phys Rev E 53:3933–3957. https://doi.org/10.1103/PhysRevE.53.3933

[7] Li Y, Yang P, Xu T, et al (2008) CFD and digital particle tracking to assess flow characteristics in the labyrinth flow path of a drip irrigation emitter. Irrig Sci 26:427–438. https://doi.org/10.1007/s00271-008-0108-1

[8] Salvador P, G., Valverde A, et al (2004) Hydraulic Flow Behaviour through an In-line Emitter Labyrinth using CFD Techniques. ASAE, St. Joseph, MI

[9] Jun Z, Wanhua Z, Yiping T, et al (2007) Numerical investigation of the clogging mechanism in labyrinth channel of the emitter. Int J Numer Meth Engng 70:1598–1612. https://doi.org/10.1002/nme.1935

[10] Rambidi NG, Yakovenchuck D (1999) Finding paths in a labyrinth based on reaction–diffusion media. Biosystems 51:67–72. https://doi.org/10.1016/S0303-2647(99)00022-2

[11] Wang H, Zhao Y, Wang J, et al (2016) Numerical simulation of flow characteristics for a labyrinth passage in a pressure valve. J Hydrodyn 28:629–636. https://doi.org/10.1016/S1001-6058(16)60667-4

[12] Rosenberg H (1983) Labyrinths in the Disc and Surfaces. The Annals of Mathematics 117:1. https://doi.org/10.2307/2006969

[13] Biggs NL, Lloyd EK, Wilson RJ (1976) Graph Theory: 1736–1936. Clarendon Press, Oxford.

[14] Dijkstra EW (1959) A note on two problems in connexion with graphs. Numer Math 1:269–271. https://doi.org/10.1007/BF01386390

[15] Nilsson N (1965) Some growth and ramification properties of certain integrals on algebraic manifolds. Ark Mat 5:463–476. https://doi.org/10.1007/BF02591142

[16] Green CC, Raphael B (1967) Research on intelligent question-answering system. Stanford Research Inst Menlo Park CA

[17] Hart PE, Nilsson NJ, Raphael B (1968) A formal basis for the heuristic determination of minimum cost paths. IEEE transactions on Systems Science and Cybernetics 4:100–107

[18] Zuse K (1972) Der Plankalkül. Gesellschaft für Mathematik und Datenverarbeitung 96–105
[19] Moore EF (1959) The shortest path through a maze. In: Proc. Int. Symp. Switching Theory, 1959. pp 285–292
[20] Lee CY (1961) An algorithm for path connections and its applications. IRE transactions on electronic computers 346–365
[21] Adamatzky A (ed) (2016) Advances in Physarum Machines. Springer International Publishing, Cham
[22] Adamatzky A (ed) (2017) Advances in Unconventional Computing-I. Springer International Publishing, Cham
[23] Adamatzky A (ed) (2017) Advances in Unconventional Computing-II. Springer International Publishing, Cham
[24] Cognato G de P, Bortolotto JW, Blazina AR, et al (2012) Y-Maze memory task in zebrafish (Danio rerio): the role of glutamatergic and cholinergic systems on the acquisition and consolidation periods. Neurobiology of learning and memory 98:321–328
[25] Aoki R, Tsuboi T, Okamoto H (2015) Y-maze avoidance: an automated and rapid associative learning paradigm in zebrafish. Neuroscience research 91:69–72
[26] Yokawa K, Baluška F (2017) Plant Roots as Excellent Pathfinders: Root Navigation Based on Plant Specific Sensory Systems and Sensorimotor Circuits. In: Adamatzky A (ed) Advances in Unconventional Computing: Volume 2: Prototypes, Models and Algorithms. Springer International Publishing, Cham, pp 677–685
[27] (2020) List of finite element software packages. Wikipedia. https://en.wikipedia.org/wiki/List_of_finite_element_software_packages

Path Finding in A Labyrinth Based on Displacement of Mechanical Analogical Model

Abstract

In this study, a mechanical analogy-based approach is presented for use in the solutions of path finding problems of labyrinths. According to this approach, by preserving the labyrinth topology, a mechanical model of structural system is constructed by transforming the labyrinth branches into truss parts connected by hinges. As a result of the mechanical analysis, the solution path is determined by examining the displacements and the final geometry of the structure. Mechanical analyzes were carried out within the scope of dynamic analysis procedures, and Finite element method (FEM) was used as a numerical method. Element type is chosen as truss elements in FE models. Since the selected element type is one of the element types that can operate with the lowest computational cost that can be used in FE analysis, it is possible to reach the results very quickly. The calculation routine can be performed with any of the general-purpose FE packages as well as it can be easily implemented by coding in any coding environment.

In order to present the flexibility of the approach utilized in solution, problem is solved for two different load cases. For both loading cases, it is presented not only the results obtained, but also the visualization of the different phases of the solution in order to make the method more understandable.

1 Introduction

Labyrinths are structures that are topologically different from mazes, and they are geometric forms that are known by human beings, built for various purposes and take place in illustrations even before their mathematical properties are defined [1, 2]. Labyrinths, which were dealt with for various reasons such as religion or aesthetics in the past, are important for their solutions today. Because it is possible to apply the solution techniques of labyrinths to many kinds of problems belonging to different fields. [3~8]*.

---

* The references presented in this study includes not only the labyrinths

Elaboration and classification of methods available for labyrinths under a certain discipline is a relatively novel subject, in comparison with the improvements in other branches of mathematics [9]. In this context, most of the methods commonly used today are discussed under Graph theory, a branch of mathematics [11-18]. For the purpose of solving these problem types, they are available in approaches that are outside the direct mathematical approaches and in which the accumulation of knowledge in different fields is handled with analogical approaches. In these studies, various behaviors and phenomena such like chemical reactions, food search of bacteria and fungi are formulized and transformed into algorithms [19~21]*. Formulations of these behaviors and phenomena are not always easy, but unanticipated benefits are the basis behind the driving force of studies carried out within this scope.

In the scope of this study oriented on solution of labyrinths, as can be observed in the study done by Altintas [22], a solution based on mechanical analogies is presented. As distinct from study of Altintas [22], explicit dynamic analysis procedure in FEM is utilized with truss elements in this approach [23], and solution is obtained by use of the displacements instead of stresses.

2 Problem definition and solution technique

The labyrinth geometry to be solved in the scope of this study is given in Fig. 1. It is required that the path connecting points A and B should be determined. The mechanical approach to be adopted is FEM and the element type is truss. Truss parts are elements that can be connected to each other with hinges. If the labyrinth geometry to be examined is used in the FE model where each piece is modeled with more than one element, it is not possible for the structural system to

---

studies but also the mazes. Although the terms maze and labyrinth are often used in the same sense in the literature, labyrinths and mazes are topologically different from each other. (See also: A. Adamatzky, Physical Maze Solvers. All Twelve Prototypes Implement 1961 Lee Algorithm). However, except several exceptions, solution techniques used for mazes are also usable for solution of labyrinths.

* Analogical approaches are of course not limited to these references, the references presented have been selected because the current studies are compiled and the previous references can be easily followed up.

keep its form under load, but this is exactly what is desired in the approach to be used. Boundary conditions to be used in solution of system are based on a pinned support and a load moving away from it. It is expected that the system having these boundary conditions will deform under loading, and in its final form, branches constituting the solution set of labyrinths or in the other words finite elements constituting the branches are aligned between these two points.

Since the structure cannot preserve its form under loading, the displacements of elements of labyrinth will be inevitably relatively high in comparison with its dimensions.

Fig. 1 Labyrinth geometry and the points to be connected with the route to be determined

For this reason, although it is not compulsory in solution process, in order to obtain a more precise solution and be able to properly include large displacements, dynamic explicit analysis procedure is used [23]. While it is appropriate to follow the dynamic analysis procedures in the problems with large displacements, it will be appropriate to reduce the additional effects of the inertial forces that are not inherent to the original problem and are inevitable to occur. The effect of these forces can be reduced by selecting values such as mass and loading speed small initially or with appropriate damping parameters to be added.

3 Analysis and results

Truss elements, one of the element types that will form the very small system stiffness matrix, have been used in the constitution of the labyrinth FE model, and it is possible that the truss elements can rotate freely around the joints due to the hinges at the connection points. Since the system is in the form of a labyrinth, truss elements cannot take the shape of rigid triangular and similar sub-bearing forms [24]. In this regard, after the structure is subjected to loading for a certain period, it is expected that the elements constituting the solution set are aligned on or near a line between points A and B. There is no need to continue the analysis after the elements forming the solution set are aligned on the line from A to B. This analyses process is quite simple and much faster than the approach in Altintas [22].

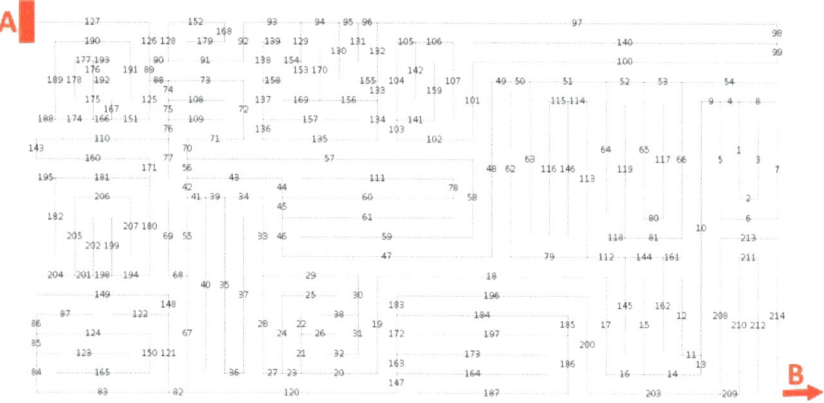

Fig. 2 FEM model of labyrinth and number of elements

Nevertheless, selection of several parameters for dynamic analysis are vital for a successful calculation process. Hitting the road by a realistic physical design and selection of parameters for minimization of inertial forces are sufficient for a successfully finalization of the process. The problem examined in the study was handled with two close approaches that differ depending on the type of loading. In the first approach, the system in gravity-free environment with supported from point A and loaded in point B is handled. In the second approach, in addition to loading in Point B, reduced gravity is applied to the system. During transformation of the original problem into a structural system that should be subjected to mechanical analyses, it is essential that boundary conditions which do not exist in original problem should not be added. In this manner, none of the mechanical simulations utilize interaction property, which has the possibility to add additional boundary conditions to problem in hand.

Exact Solutions of Shortest-Path Problems Based on Mechanical Analogies

Two approaches can be used for solution, and truss elements constituting the set of solution can easily be distinguished from others algorithmically or visually. The moment solution set is visually or algorithmically detectable, there is no need to continue the simulation.

3.1 Solution for system subjected to concentrated force

For determination of path unifying points A and B in labyrinth system, a pinned support is constituted in Point A, and point B is loaded by a force drifting apart from point A. The variation of geometric form of labyrinth system composed of truss elements will be traced. In this scope, a force, which the magnitude is linearly increased from 0 to 250 N in 25 seconds will be applied to Point B, the solution set will be determined considering the behavior of labyrinth under these boundary conditions as a basis. The rectangular labyrinth has dimensions of 780x380 mm^2 in plain view, each branch has a cross-sectional area of 10mm^2 and unit weight is assumed as 5000kg/m^3. Elastic modulus, Poisson's ratio and mass proportional damping ratio of homogenous material constituting the labyrinth are 200GPa, 0.3 and 0.25, respectively. The damping parameter is accepted as a larger value than usual or realistic values in order to immediately decrease the effects of inertial forces and vibrations. The behavior of the system under load is presented in Fig. 3 for various time sections. Academic and commercial FEA software is able to cope with problems with several million nodes even in personal computers, in this context, the solution was reached relatively quickly using the simulation of the labyrinth system consisting of 214 truss elements. The structural form of the labyrinth in Fig. 3 has changed rapidly from the moment the load was first loaded and the solution set appeared prominently in the immediate vicinity of the line between the two points as 127, 126, 89, 88, 73, 72, 71, 70, 56, 42, 55, 67, 120, 147, 163, 172, 183, 196, 200, 203, 208, 213, 214. Colors arising during analyses belong to Von-Mises stresses, and the colors of branches in solution set in final state are different those of the remaining system. During the analysis, the stresses, produced by the effect of inertial forces, occurring in branches other than the solution set have finally disappeared. In mechanical simulation of labyrinth problem, only tensile stresses are observed, therefore, it is understood that converging to the solution is not only possible by use of the final shape of the system due to displacements, utilization of stresses are also an option here. However, in order to

Exact Solutions of Shortest-Path Problems Based on Mechanical Analogies

make an assessment based on stresses, inertial forces should be attenuated and oscillations should be over.

Initial Geometry

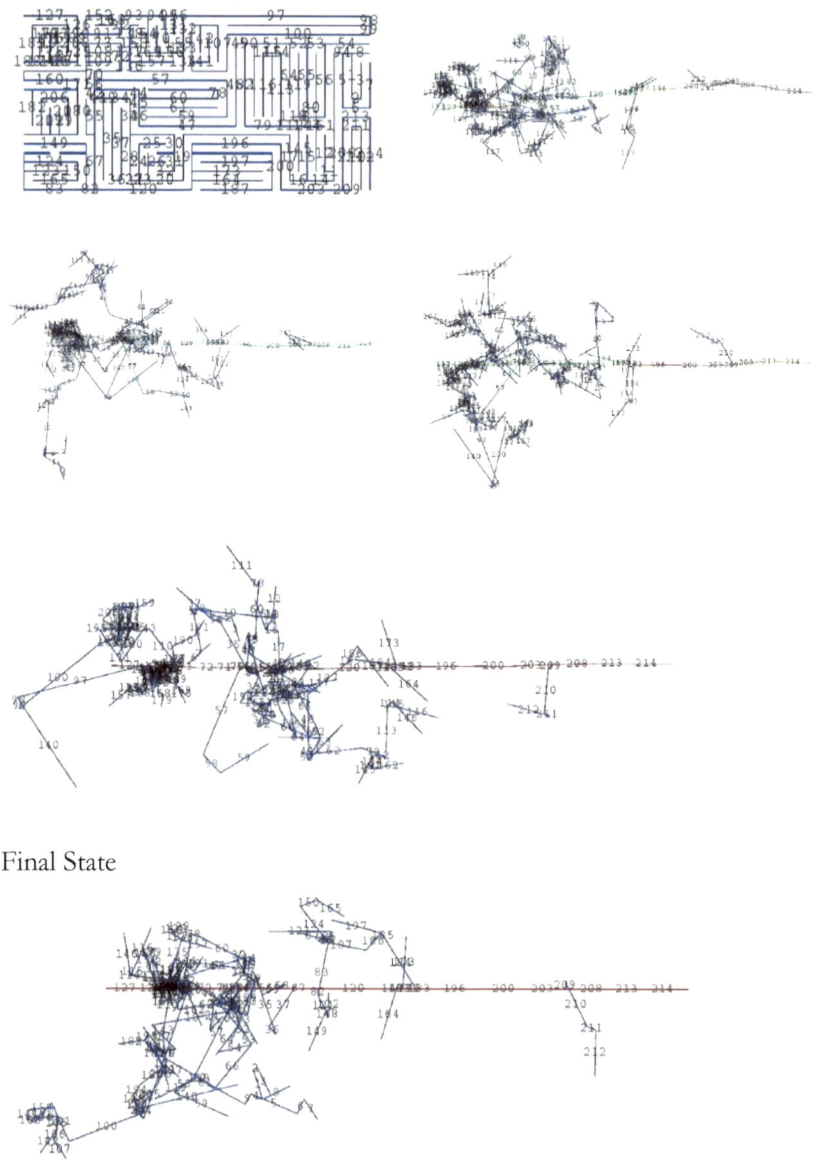

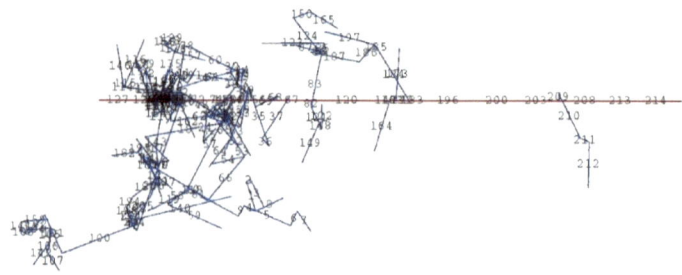

Final State

Fig. 3 Variation of appearance of structural system in time domain (for concentrated force)

3.2 Solution for tension force and gravity

The aim of the solution made in this section is to show that approaches with different mechanical scenarios can be used in obtaining the solution set as well as to present a more visually distinctive form of the solution set. In this solution, which is presented as the second approach for the solution of the labyrinth problem, as distinct from the first approach, the reduced gravity on the system was applied as 1000mm/s2. In this manner, it is planned that branches excluded by the solution set and eventually the elements are easily discernible.

In Fig. 4, the outputs of simulation in this scope are presented and the set of solution is visually observed clearly in the bottom image. The inertial movements arising from loading at first sight are attenuated after the set of solution attains its final shape. The final shape of the labyrinth system has been achieved quite quickly due to suitable damping and gravity values. In the final case, similar to the first approach, the solution set is located close to the hypothetical line between the support and load points. Other branches or finite elements outside the solution set are located on the lines in the direction of gravity. Colors of stresses at elements composing the set of solutions are different from those of remaining elements. In the second approach, the stress values cannot be used safely in determining the solution set even if the structural form has reached its final position. Because, due to gravity, there is no any obstacles for occurrence of the greatest stress values on the elements outside of the solution set. On the other hand, whether we apply the gravity value as a reduced value or not, it is impossible to observe a complete set of solution on a completely straight path. However, as long as the theory of small deformations is valid, it should be expected that in the system that has reached its final form, the top of the lines connecting the two points is the solution set.

While the visibility of the solution set observed in the second example is more successful than the approach in the first example, it should be considered that the approach in the first example can be more reliable.

Exact Solutions of Shortest-Path Problems Based on Mechanical Analogies

Initial state

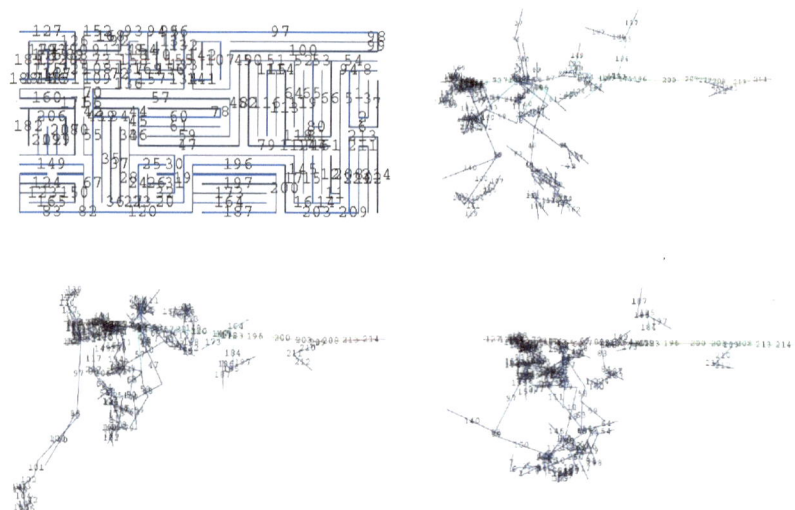

Final State

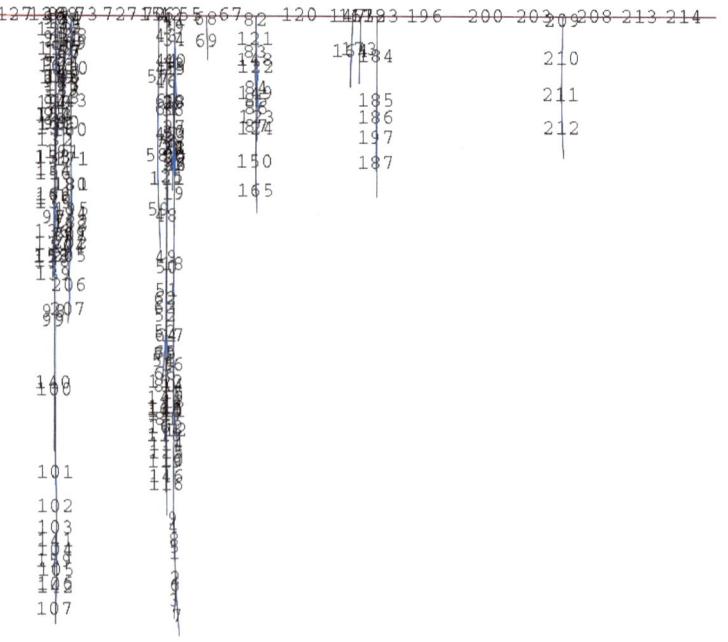

Fig. 4 Variation of appearance of structural system in time domain (for concentrated force and gravity)

4 Conclusions

Mechanical based solutions not only provide an abundance of scenarios, at the same time, they have a vast amount of options, to reshape these scenarios pertinent to need for different computer configurations. In this scope, this study presents solutions by use of two similar approaches. It should be emphasized that small differences in approaches lead to great differences in structural behaviors and determining factors influencing on results.

The first solution is based on determination of path unifying points A and B in labyrinth system, a pinned support is constituted in point A, and point B is loaded by a force drifting apart from point A. Results revealed that, path connecting the two points can be detected by use of displacements and stresses. In the second approach, differently from the first approach, the reliable solution can only be obtained by observing displacements of system under the effect of gravity. The two approaches enable the user to obtain the solution set based on displacements, however, solution based on stresses are not reliable for second approach. Since the solution set can be obtained precisely by evaluating the displacements in connection with the geometric form of the structure, the solutions can also be obtained in some methods other than the FEM. In this scope, structure of labyrinth can also be constituted using interconnected rigid pieces, such like interconnected truss elements in FE analyses, and with solvers working within the framework of rigid body dynamics (RBD) rules. RBD based solutions within this framework can be more advantageous in comparison with FEM, taking the computational cost as a dependent parameter. But, in addition to vast number of parameters of structural models, FEM as numerical analysis method provides flexibility and adaptability by additional parameters stemming from its own solution procedures and types of elements. For instance, it is almost impossible to apply an RBD based solution of the labyrinth modeled as a single solid, akin to the mechanical approach provided in [22]*. Considering the computational costs, the element type used in this study is more advantageous in comparison with that used in Altintas [22]. However,

---

* However, it may be possible to obtain approximately stress etc. values indirectly by various algorithms by considering the forces and their positions to the connection points of the elements.

differences in mechanical model and the FEM procedure* used in solutions influence the computational cost of the approaches, and determination of the solution way in accordance with the problem type is not a hard task. The solutions of the mechanical models used in the scope of the presented approaches can easily be obtained by commonly used FE solvers or scripts coded based on the approaches.

References

[1] Matthews WH (1922) Mazes and Labyrinths. The Ballantyne Press Spottiswoode, Ballantyne & Co. Ltd., Colchester, London & Eton
[2] Doob PR (1992) The idea of the labyrinth: from classical antiquity through the Middle Ages, First publ. 1990. First print., Cornell Paperbacks. Cornell Univ. Pr, Ithaca u.a
[3] Nakagaki T, Yamada H, Tóth Á (2001) Path finding by tube morphogenesis in an amoeboid organism. Biophysical Chemistry 92:47–52. https://doi.org/10.1016/S0301-4622(01)00179-X
[4] Podsędkowski L, Nowakowski J, Idzikowski M, Vizvary I (2001) A new solution for path planning in partially known or unknown environment for nonholonomic mobile robots. Robotics and Autonomous Systems 34:145–152. https://doi.org/10.1016/S0921-8890(00)00118-4
[5] Adamatzky A, de Lacy Costello B (2002) Collision-free path planning in the Belousov-Zhabotinsky medium assisted by a cellular automaton. Naturwissenschaften 89:474–478. https://doi.org/10.1007/s00114-002-0363-6
[6] Zhang L, Chiaradia A, Zhuang Y (2013) In The Intelligibility Maze of Space Syntax–a Space Syntax Analysis of Toy Models, Mazes and Labyrinths
[7] Vourkas I, Sirakoulis GCh (2016) Networks of Memristors and Memristive Components. In: Memristor-Based Nanoelectronic Computing Circuits and Architectures. Springer International Publishing, Cham, pp 173–198
[8] Adamatzky A (2017) Physical maze solvers. All twelve prototypes implement 1961 Lee algorithm. In: Emergent computation. Springer, pp 489–504

---

* Static, dynamic explicit, dynamic implicit etc.

[9] Biggs NL, Lloyd EK, Wilson RJ (1976) Graph Theory: 1736–1936. Clarendon Press, Oxford
[10] Shimbel A (1954) Structure in communication nets. In: Proceedings of the symposium on information networks. Polytechnic Institute of Brooklyn, pp 119–203
[11] Moore EF (1959) The shortest path through a maze. In: Proc. Int. Symp. Switching Theory, 1959. pp 285–292
[12] Bellman R (1958) On a routing problem. Quarterly of applied mathematics 16:87–90
[13] Dijkstra EW (1959) A note on two problems in connexion with graphs. Numerische mathematik 1:269–271
[14] Nilsson N (1965) Some growth and ramification properties of certain integrals on algebraic manifolds. Arkiv för matematik 5:463–476
[15] Green CC, Raphael B (1967) Research on intelligent question-answering system. Stanford Research Inst Menlo Park CA
[16] Hart PE, Nilsson NJ, Raphael B (1968) A formal basis for the heuristic determination of minimum cost paths. IEEE transactions on Systems Science and Cybernetics 4:100–107
[17] Zuse K (1972) Der Plankalkül. Gesellschaft für Mathematik und Datenverarbeitung 96–105
[18] Lee CY (1961) An algorithm for path connections and its applications. IRE transactions on electronic computers 346–365
[19] Adamatzky A (ed) (2016) Advances in Physarum Machines. Springer International Publishing, Cham
[20] Adamatzky A (ed) (2017) Advances in Unconventional Computing-I. Springer International Publishing, Cham
[21] Adamatzky A (ed) (2017) Advances in Unconventional Computing-II. Springer International Publishing, Cham
[22] Altintas G (2020) Path Finding in a Labyrinth Based on Stress Distribution of Mechanical Analogical Model. In: Exact Solutions of Shortest-Path Problems Based on Mechanical Analogies. Amazon, pp 15-28
[23] Smith M (2009) ABAQUS/Standard User's Manual, Version 6.9. Dassault Systèmes Simulia Corp, United States
[24] Altintas G (2020) Exact Solution of The Shortest Path in a Maze Based on Mechanical Analogies and Considerations. In: Exact Solutions of Shortest-Path Problems Based on Mechanical Analogies. Amazon, pp 40-71

Exact Solution of The Shortest Path in a Maze Based on Mechanical Analogies and Considerations

Abstract

The solution of a maze is based on determining the shortest path from the all possible solutions between the entry and exit points. Methods used in solution of mazes not only determine the shortest paths connecting two points in a geometrical form, but provide methods for solutions of all associated optimization problems which can be transformed into this form. In this study, solution methods based on mechanical analogies which can be used in exact solutions of mazes are presented.

With the proposed approaches in this study, it is shown that the exact solutions of maze problems in two and three-dimensional environments are obtained. According to the approaches, solution methods include transformation of maze geometries, and boundary conditions into structural systems and obtainment of numerical solutions by using finite element method (FEM). Efficiency of presented approaches are comparatively investigated for different cases and information ensuring success of solutions are enlisted.

It is possible to realize the proposed approaches even with game engines where simulations can be made based on the principles of Rigid Body dynamics (RBD), which use less parameters compared to the FEM, and presented in the solution in this context. Due to the fact that both FEM or RBD based solution techniques can be used in solutions, they provide a wide pool of options that can be used in the solution depending on the computation capacity and many other parameters.

1 Introduction

Path finding and shortest path problems in connection with labyrinths and mazes are among the well-established issues of the literature of Graph theory and optimization algorithms [1, 2]. In this framework, optimization problems that can be transformed into maze problems can be solved by a similar approach to that of Altintas [3], focused on solution of mazes in this study.

In this scope, two mazes are transformed into structural systems by mechanical analogy and solution is obtained by FEM. In order to demonstrate the flexibility of approaches, solutions of 2D and 3D models have been made, and the approaches used due to the absence of situations requiring mechanical interactions in the original optimization problems within the scope of the examination can be used not only in geometrically but also in topologically different maze solutions[4-7]. Another reason for the solution of problem solving with three-dimensional and curvilinear path is to address the situations that need attention in the usage of curvilinear paths originated from real life geometries and to show the transformation of 3D models into 2D models.

In this study, Graph theory terms are often used in definitions of mazes and transformed models based on mechanical analogy. However, the most important point is, apart from proposed approaches, when using graphs in the scope of Graph theory, there is no need to show the lines scaled, based on their costs. For application of the approach proposed, the branch lengths of mazes, transformed into models in the scope of mechanical analogy should be in accordance with costs of original problem. Although this situation seems to be a constraint at first, it is possible to transfer the costs normally not possible to be shown with a scaled single straight line to mechanical analogies, with the virtual node adding technique presented in this study. The origin of the more robust method used in Altintas [8] is very similar to the technique presented in this study. FEM was mainly used for the solution of the problems as the numerical method, and one of the examples was carried out using Unity3D game engine [9] with an RBD based approach.

The cases presented in the scope of this study are basic examples providing ease in understanding the approach as well as the subject. In additional solutions presented, assumptions and parameters affecting the results are investigated, and several warnings and suggestions are provided.

1.1 Brief information about approaches based on mechanical analogies

A series of simple operations and assumptions are necessary to realize the proposed approaches. The edges of the graphs, which are not scaled in their graph representation, are scaled in proportion to their costs and are converted into branches of mazes by preserving the topology of the problem. In the next step, the structure in the maze form is divided into structural elements so that their solution can be performed by FEM or RBD based methods. At the last step, by making the definitions of material properties and boundary conditions, the system is analyzed in accordance with the mechanical analogy and the results are obtained. Giving a little more detail about the approaches briefly described above and the topics mentioned in the study will provide a better evaluation of the presented approaches.

FEM was predominantly used in analysis of the mechanical models and branches belonging to mazes were to be modelled by truss elements as demonstrated by Altintas [3]. In order to solve a shortest path problem that can be converted into a maze form by mechanical analogy, the costs and boundary conditions of the original problem should be transferred to the mechanical structure appropriately. As distinct from labyrinths, the lengths of the maze branches in the mechanical model must be compatible with the costs. Because truss lengths in mechanical systems should not be unscaled, as in most graphs edges (lines).

Here, one question may arise: "Is it possible to always create a mechanical structure depending on the actual problem costs?". Because it is not possible to transfer each cost of the real problem to the maze system where the structural solution will be made, with a single straight-line segment, except for special cases. However, when such a situation is encountered, virtual nodes can be additionally included in the system to obtain a cost length by addition of linear truss elements. A single node between two primary nodes adequate for this aim. Mechanical analogies are not completely independent from the numerical solution technique to be used most of the time. As an example, in a numerical analysis by using FEM, if the branches forming the maze should freely rotate around the connection nodes, certain element types such like truss elements should be utilized. Of course, these conditions can also be realized with appropriate combinations of various element types with various degrees of freedom. Theoretically, there is no upper limit for the number of elements to be converted from branches. However, it is beneficial to keep it to a minimum as the increase in the number of elements will cause the computational cost to increase. On the other hand, there is a lower limit on the number of elements, the relevant details are discussed in detail in Sect.

3. Entry and exit points of the maze are essential boundary conditions used in analysis of the system. While one of these nodes is transformed into a pinned support, the other is subjected to a load which is loaded in an opposite direction from the support, as in Altintas [3]. However, as distinct from labyrinths, there are more than one path connecting the entrance and exit points in the mazes. For this reason, there will be some differences in determining the solution set of the mazes, in the evaluation of displacements, final positions and stresses. It is possible to obtain the solutions depending on final geometric form and displacements even by RBD based movement engines. In this scope, a solution performed by use of game engine Unity3D is also presented. Most of the cautions for geometric definitions in FEM are also valid for application of optimization problems in game engines by the proposed approaches. Since the game engines commonly used in practice are not usually produced for numerical mechanic analyzes, it will be useful to take care of the warnings in the relevant section regarding the use of RBD based solutions. None of the approaches used in study includes use of interaction. Because, the interaction of structural system elements used in application of mechanical analogy results in formation of additional boundary conditions that were not exist in original problem. For this reason, the interaction feature should not be used unless a specific boundary condition needs to be defined. Thanks to the disuse of the interaction feature and the proposed approach, non-planar mazes and problems that are actually three-dimensional can be handled and examined as two-dimensional problems without encountering any problems.

In this study, examples are prepared not only for application of the approaches used, at the same time, additional solutions were carried out to find the effects of several assumptions and options on results, and provide a better understanding of the solution abilities of approaches in the framework of mechanical analogies.

2 Shortest path solution of the maze

The graph* that appears with the scaled drawing of the branches

---

* Depending on the costs between the vertexes of shortest path problems, scaled straight lines come together and create a structural form with continuity is generally used in certain areas such as problems using real distances. However, the approaches presented in this study can also be used between the vertexes of the example shortest path problem is presented in Fig. 1 and the numbers of the vertex and lines are shown in the same graph. In the problem, it is desired to determine the shortest path between points 11 and 14, and the system to be used in the solution has 31 lines and 14 vertexes.

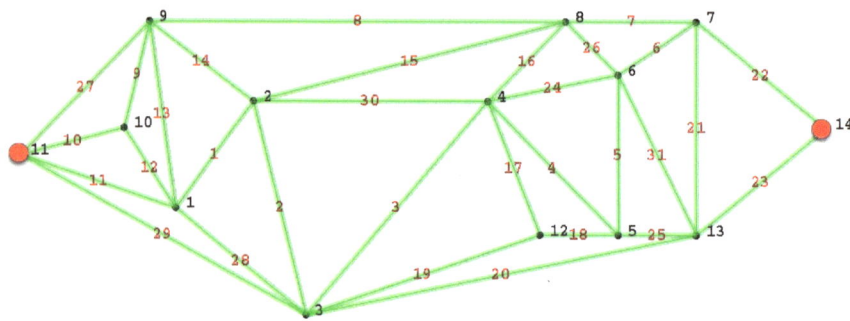

Line	Length	Line	Length	Line	Length	Line	Length	Line	Length
1	25	8	80	15	61.8466	22	32.0156	29	62.6498
2	41.2311	9	20.6155	16	21.2132	23	32.0156	30	45
3	53.1507	10	20.6155	17	26.9258	24	25.4951	31	33.541
4	35.3553	11	31.6228	18	15	25	15		
5	30	12	18.0278	19	47.4342	26	14.1421		
6	18.0278	13	35.3553	20	76.4853	27	35.3553		
7	25	14	25	21	40	28	32.0156		

Fig. 1 Scaled plan of shortest path problem and geometrical details (Units SI-mm)

As applied in solution of labyrinth problem [3], the path is composed of imaginary lines which are located between the points of application of load which is moving away from the support and the support itself. With the similar approach, the load and support cond (a) ed for the mechanical model of maze are presented in Fig. 2a.

---

in the solution of problems where it is not possible to create structural forms using straigth lines connecting the vertexes on the plane. Some additional approaches need to be made for this achievement, and these approaches are addressed in Sect. 3.

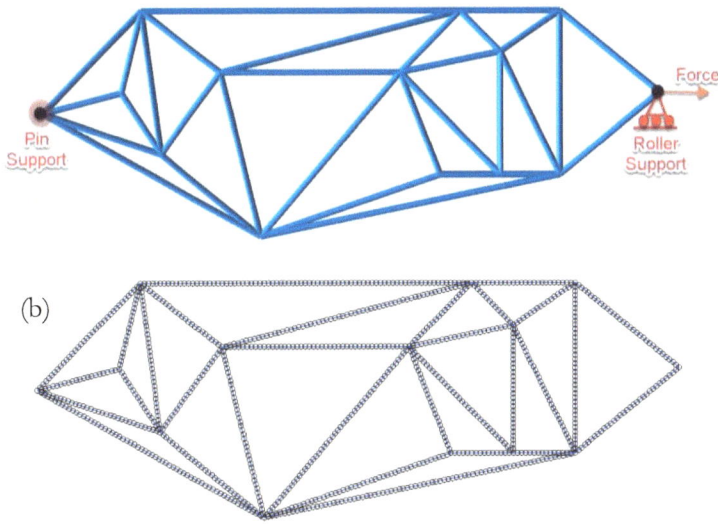

Fig. 2 a Loading and supports of mechanically analogous model. **b** FE mesh for approximately 1mm element size

In Fig. 2, the boundary conditions, and the mesh of the examined system consisting of truss elements of approximately 1 mm length are presented. The number of elements used in system is far above of the minimum requirements. Nevertheless, in the solution of real-life truss systems in civil or mechanical engineering field, generally, every part between two nodes would essentially be defined as a single truss element in FE model. However, this type of approach is not suitable for the solution of maze problems as it may cause situations such as additional constraints that do not exist in the real shortest path problem even if it includes the least number of elements. In such cases, there are simple approaches that must be followed carefully to determine the minimum number of elements (see the Sect. 3). For this reason, in this section, a truss piece between two points is modeled with multiple truss elements that can easily rotate around the connection points. Thus, in accordance with the essence of the approach used, the interconnected parts can move freely like a rope. When a system that can move in this way is subjected to loading, it is not difficult to expect that the shortest path between the support and the load will be the path closest to the line connecting these two points. A system that behaves in this way also allows the results to be visualized and understandable in a natural way. However, it should not be forgotten that, as in this example, the use of more elements than needed causes the results of the problems that can be solved in a much shorter time to be achieved in a longer time. The computational cost due to excessive number of elements can be reduced by selecting an appropriate number of elements (see the Sect. 3).

Table 1 includes information about the system under consideration by means of loading, materials, parameters regarding FE mesh. If static analysis procedure was opted in solution of problem, there was no need for optional parameters. In the dynamic analysis procedure, some of these parameters may not be used or the manipulations of their values with reasonable value will not change the final solution set of the problem, but it may increase the time needed for the solution*. However, it is not required to be used in this problem, the purpose of using some parameters included in analysis such as gravity is to familiarize oneself with the adaptation and use of mechanical analogies to different problem types when necessary.

Table 1 Load, material and mesh definitions (Units are SI-mm)

Load and material definitions		Optional Variables and Values	
Force	100	Selected Gravity Value†	9810
Young Modulus	200000	Rayleigh Damping (Alpha)‡	10
Poisson's Ratio	0.3	Density §	8e-9
Cross section area of truss elements	2		
Mesh Properties			
Number of Nodes	1057	Element Type	2D Truss Line
Number of Elements	1074	Approximate Element Size	1

---

* The use of the static analysis procedure in systems with large displacements may result in poor performance in achieving results or no results at all.

† There is no need to use Gravity in the analysis, it is only used to visually differentiate the result.

‡ The reason that the mass proportional damping value was chosen as a value much higher than the realistic values is because it reduces the simulation time required to determine the solution set. Thanks to this value, inertial forces are absorbed very quickly.

§ The reason for using density is that the analysis is desired to be performed as a dynamic explicit. If the solution will be made with static procedures, the density parameter may not be used. However, in cases where large displacements are the subject, the statistic analysis procedure should be abandoned.

In solution of problem, FEM was used to assess the behavior of the system under loading by use of dynamic explicit method. Time dependent behavior of the system obtained by the use of FEM was presented in Fig. 3. Utilizing the values in Table 1, simulation was proceeded until the elements outside the solution were completely separation from the solution set, under the effects of gravity. As in the example where visual intelligibility is prioritized, the solution procedure can be terminated as soon as a viable set of solution notices visually.

Of course, the separation of the solution set from other elements could be determined in the algorithmic way, and this is already the method that should be followed in large and comprehensive shortest path problems related to real life. However, in order to better understand the approaches in this study, this kind of high visual expression style was preferred.

The set of solution of the system under consideration was determined in half of the time that system reached its final form. In fact, when the set of solution is apparently observed or algorithmically determined, there is no need to continue the solution. Because, as this system of elastic structure continues to increase deformations under the effect of increasing force, the elements that should not normally be in the solution set begin to be positioned close to or above the hypothetical line connecting the two points with boundary conditions.*

Based on the displacements from the results, the shortest path connecting nodes 11 and 14 is composed of lines 29-20-23, based on the graph numeration in Fig. 1.

The colors in Fig. 3 vary depending on the stresses (Von-Mises) that occur during the analysis, and the reason why this coloring is included in the presentation is that the results of the analysis can be interpreted not only in terms of displacements but also in terms of stresses.

---

* Please see Sect. 3.2 and Fig. 7b

Initial State

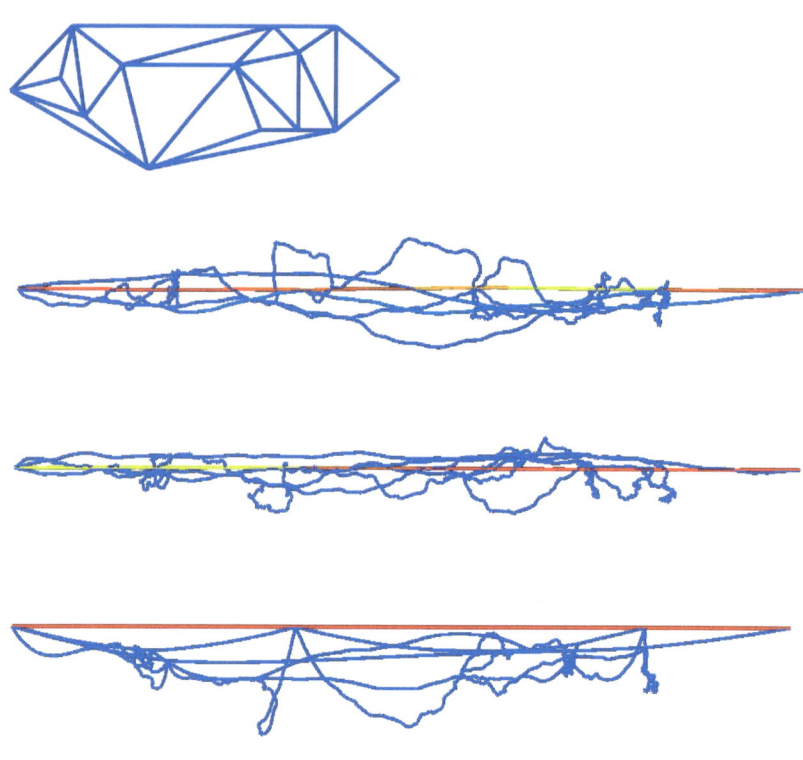

Final State

Fig. 3 Variation of appearance of structural system in time domain

When the results in Fig. 3 are examined, it is immediately noticeable that the colors of the stresses of the elements in the solution set are different from the elements other than the solution set. Although the same solution set was obtained in terms of stress and displacement-based evaluations in the examined problem, it is not reliable to identify the solution set based on the stresses, except in special cases, in the mazes. For example, in a system with a different connection structure, there is always possibility that the loads resulting from the weight of the elements are concentrated in the elements not in the solution set and the stress values exceed the ones in the solution set. For this reason, evaluating the displacements and the positions of the elements in the system will yield reliable results.

Information in succeeding section regarding to matters influencing on the success of the results and important warnings about wrong mechanical analogies are of utmost importance for correct use of approaches presented in this study.

3 Considerations on maze solutions by using mechanical analogy and FEM

3.1 Mesh resolution issue

In this section, a new FE model is created by using fewer number of nodes and elements for the solution of maze solved in Sect. 2. In this way, it is expected that the possible solution of system will be realized with a lower computational cost and reach the results more quickly.

However, in systems with reduced number of elements or improper configuration without following certain rules, it may not be possible to obtain the solution quickly, even the solution set of the shortest path problem. In this context, information will be given not only on the predictable relationship between the number of elements and the computational cost of the solution, but also how to determine the minimum number of elements that can be used essentially.

The analyzes made in this section are designed by focusing on the possible use errors of mechanical analogies and numerical methods in connection with these analogies. In this context, in the solution presented in Fig. 4, the mechanical behavior that would occur if less elements were used than needed was exhibited.

Triangles shown in pink in Fig 4a were not spoiled even at the end of analysis as shown in Fig. 4b. The truss pieces forming the preserved form of triangles are represented by a single truss element and these triangles should be accepted to be stable structures, hindering the correct arrangement of set of solution between the entry and exit nodes.

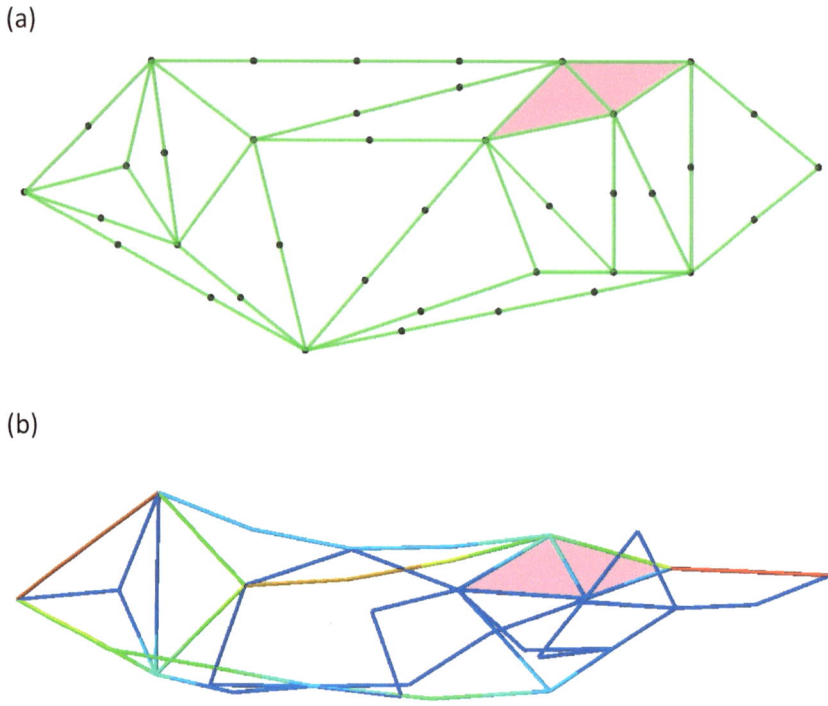

Fig. 4. Model transferred to mechanical analogy using insufficient number of elements and the solution. **a** FE Nodes on Maze. **b** Final form of analyzed maze

Stable triangular structures cause various types of limitations in displacements, leading to formation of barriers on behavior necessary for the solution. In the FE model of the system with this structural geometry, which maintains its form under load, it will not be possible to reach the solution with the proposed approach if no additional nodes are added to the FE mesh. Because in such a system, additional constraints that do not actually exist are incorrectly included in the model.

In the approach used, the creation of the model based on mechanical analogy and the FE mesh are not very independent from each other. FE programs, which are widely used, can be used to insert nodes. These programs can use different strategies in the stage of mesh generation or alteration depending on many variables such as topology, geometry, problem, and analysis types [10-16].

However, the needs of the approach discussed in this section are a bit different and it is important to check whether the automatic mesh generation process provides these needs. While it is not the most appropriate approach to minimize the number of elements, it may be appropriate as a practical approach to represent each line of the original problem by two truss elements in FE model. It should be emphasized that this is a condition that may not be always provided by commonly used FEM software.

In spite of this, a more common approach is the determination mesh structure by an average element size for use in different parts of the system and a parameter which gradually changes this size, when necessary.

In this study, as can be observed in some of the problems solved, the approximate size of the element can be accepted as the dimension which can divide the smallest truss part into two truss elements. However, if there are large distance differences between the points, the advantage of this approach may turn into a disadvantage.

Because of the large differences between the truss parts, the global element size to be adjusted according to the smallest truss part size will cause long truss parts to be divided into more elements than needed. This situation can be observed easily in Fig. 5a, due to the determination of approximate global size based on one of the small lines of system, which has been used to prevent the formation of rigid triangular structures, long parts have been divided into more elements than necessary. This means that the FE mesh is not at its optimum state.

However, in order to obtain the solution presented in Fig. 3, the result was reached with only 94 elements, which are much less than the 1057 elements used in analysis of the system in Fig. 3. As in Fig. 5a, the single undivided truss is not sufficient for formation of rigid triangular elements, before it causes formation of rigid triangular structures, it is possible to remove many points to carry put calculations with lower number of nodes.

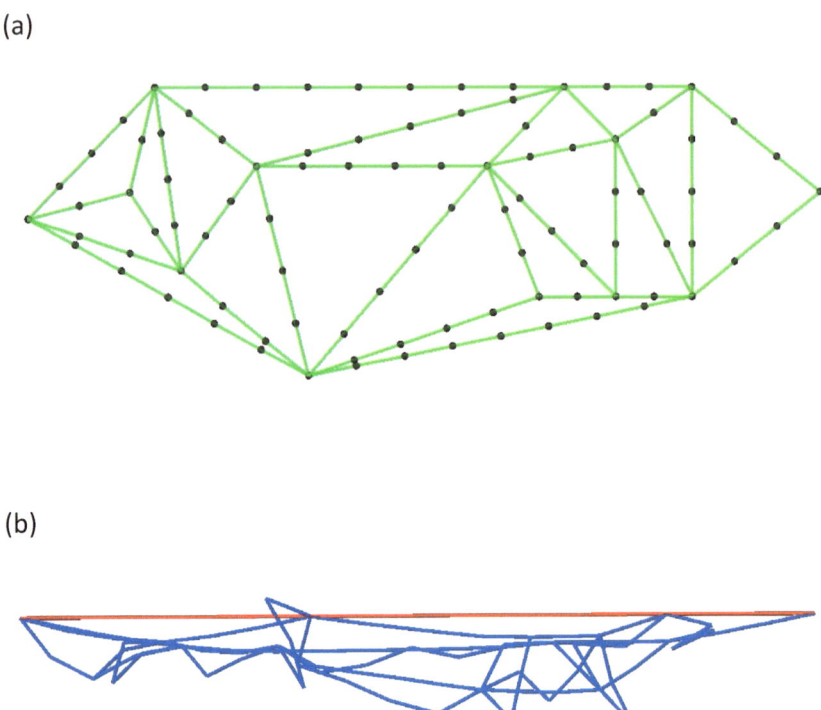

Fig. 5 a FE model free of rigid triangular elements. **b** Result of the solution

The final form of the system shown in Fig. 5b, where the solution set of the problem can be clearly seen, reaching the solution required less computational cost and reaching the results faster since it was obtained with fewer elements than the model shown in Fig. 3.

3.2 Considerations on selection of loads, materials and section properties

It will be useful to briefly touch upon some issues that should be considered about the selection of parameters such as load, material and cross-sectional properties used solutions of shortest path problems using mechanical analogies.

First of all, it should be noted that the mechanical analogy is a tool in shortest path problems and should be constituted in accordance with boundary conditions of the original shortest path problem. It is very important for working with lower computational costs that the simulated mechanical behaviors remain within the linear behavior limits in terms of geometrical and material as much as possible without exhibiting excessive stress and deformations.

In solution of shortest path problems, apart from the one presented in this study, many types of mechanics-based analysis can be used. The main approach used in this study is based on the idea that the shortest path between two points appears as a line connecting these two points, when two points are separating from each other in an appropriate amount.

The effect of gravity, which is used for the clarity of the results, is not one of the main elements of the solution approach. Moreover, if necessary, attention is not paid, obtaining a wrong solution set is probable. In this context, it would be appropriate to evaluate the solutions of the sample problem made for the cases where the gravity was and did not.

The result of the system shown in Fig. 6a * is presented in Fig. 6b for the case where there is a gravitational effect in addition to the concentrated load, and the solution made for the case of only concentrated load is presented in Fig. 6c.

Although the results obtained are exactly the same, the contribution of the gravity effect used to the intelligibility of the results is obvious. However, there is no need for the use of gravity since the results of the comprehensive and large problems of the real world will have to be processed algorithmically, not visually.

---

* The problem solved in this section is the same as the problem solved in Sect. 2. Roller support, which is added to the region where the load is located, has been added to make the behavior more stable and has no effect on the solution set.

(a)

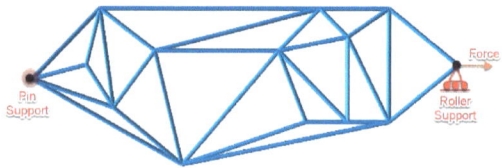

(b)

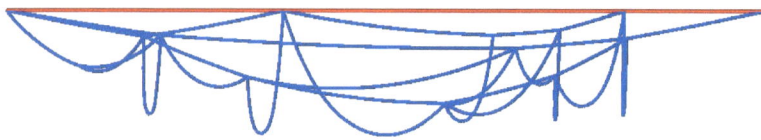

(c)

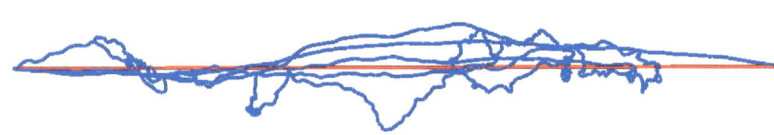

Fig. 6 Solutions of maze system. **a** original form of truss system, **b** with gravity, **c** without gravity

It would be useful to mention one more issue regarding the use of the effect of gravity. There should be a ratio between the gravity and the tension force within certain limits. The dominance of any force as in Fig. 7 can make the solution of the problem difficult or impossible. However, it is not an appropriate approach to leave anyone the essence of the problem and to make predictions about the complex issues of mechanics. In order not to cause such a situation, it will be appropriate to examine the situation in question a little before touching on what needs to be done. To discuss the problems arising from arbitrary selection of the numerical values of parameters, it would be better to select unrealistic or inappropriate values in terms of nature of problem. In this scope, Fig. 7 presents the final structural forms of previously solved problem, this time the solution of system was obtained by instruction of inappropriate values of tensile force and gravity. In the first solution the system is exposed to tensile forces and gravity at the same time, and the outcomes reveal a final form which is dominantly controlled by gravity (Fig. 7a). Differently from the solution set shown in Fig. 6, the solution did not occur in a linear or near-straight line between the two points. Although it is thought that the solution set may sometimes emerge as the top lines, it does not always allow the results to be obtained with stable reliability due to situations such as due to the gravity effect and the connection topology. Similar situations may arise with improper selections of values for parameters such as cross section geometry and modulus of elasticity that have an impact on rigidity.

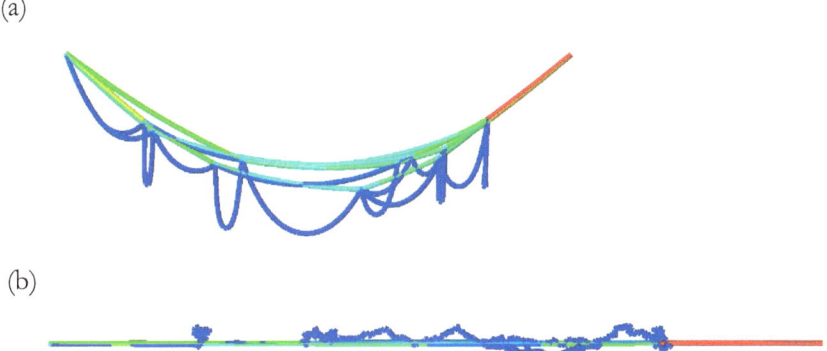

Fig. 7 Results of various scenarios. **a** case dominated by gravity, **b** case dominated by excessive tension force

In the second solution, a result indicating that the tension force is at a value that can cause excessive deformations in the system has been reached (Fig. 7b). At first glance, although the elements belonging to the solution set are considered on the linear line between the support and loading points, as in the results in Fig. 6, there are also elements not included in the solution set on the line due to excessive deformations. Nevertheless, in second solution, apart from first one, if the analysis is terminated just after the formation of first linear path before the outbreak of excessive deformations, a correct set of solution can still be obtained. Although there are problems that are easy to deal with, secondary parameters involved in different stages of analysis may also create barriers to reach to a reasonable solution. As an example, when the dynamic analysis is concerned, improper loading speeds may cause additional accelerations and inertial effects. In such a case, if a random damping value is used, the time it takes to dampen inertial forces and determine the solution set can be considerably longer. Large damping values that can be supported by numerical-analyzers, even if not realistic, or reduced mass density values that do not interfere with the dynamic analysis procedure, can practically reduce the effects of inertial terms and speed up analysis.

FEM can use time dependent displacements as loadings or boundary conditions. In this regard, above mentioned adverse effects of rate of loading can be decreased. Instead of application of loads, which its effects on rate of displacement cannot be foreseen, effect of displacement which is applied to the same point is predictable and reliable. In the examples following this part of the study, this approach was used, and the simulation was stopped at the first time when the solution set became evident in the analyzes, and the results were obtained. This approach can also easily be applied to RBD based approaches.

3.3 Interaction unused property

It was previously noted that in current approach, it is advised that the physical behaviors should be modeled in accordance with the real-life behavior, which is important for the accuracy of the results. In this regard, many issues regarding assumptions in geometry, loading, materials and FEM are given. However, if there is no particular case for the proposed approach to be applied properly, there is a behavior that should be neglected. Simply defined as interaction, this includes taking into consideration the structure's contact with itself or other bodies in the analyses. This feature, which must be included in the analysis to make physical behavior more realistic, should never be used in this proposed approach for the solution of shortest path problems unless there are any special requirement*. No interaction feature was used in any of the sample problems presented in the study. The main reason for not using the interaction is that additional boundary

---

* There is a possibility of utilization of interaction behavior, in this case, any boundary condition can be defined using mechanical interaction by including in original shortest path problem. Logical gates or complex structures can entail this.

conditions should not be included in the simulation, except for the boundary conditions of the original shortest path problem.

It will be useful to look at Fig. 8 so that people who are not familiar with mechanical behaviors and definitions can also understand the interaction behavior.

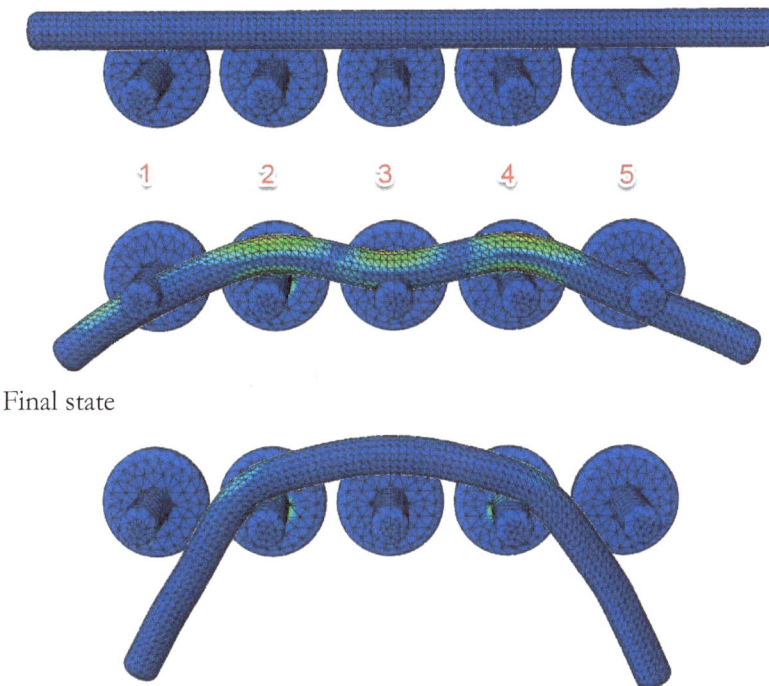

Fig. 8 Simple interaction behavior

In Fig. 8, the physical behavior of a softer body falling over five short beams is simulated using the FEM. Interaction property is defined among falling body and beams 2 and 4. Observing the ordered figures showing the behavior from start to final form, the effect of interaction is clear. Although the object interacted with beams 2 and 4, it did not interact with beams 1, 3 and 5 at all. Absence of interaction is one of the main issues of the approach, which do not exist in original shortest path problem. None of the conditions which is not belong to in original shortest path problem should be included within the problem. One of the important advantages of not using the feature of the interaction in the approach used in the study is that it allows the shortest path problems to be identified and solved in a planar manner.

3.4 FE definitions of curved lines

In addition to the considerations and suggestions in other subsections of Sect. 3, additional considerations with regards to FEM and mesh are presented. First of this is cautions is use of curvilinear forms. As a general warning, during transformation of curved to linear forms (transferring the problem into mechanical model) unexpected differences can be occurred. In Fig. 9, paths in different forms and their transformation to finite elements in terms of linear trusses are presented.

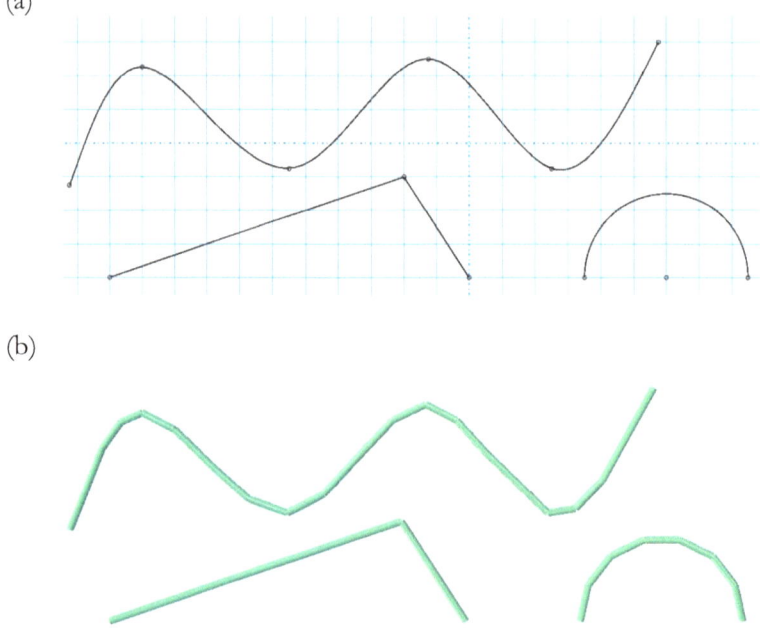

Fig. 9 a analog distances (costs of original problem), **b** distances discretized by linear elements

Whether the curvilinear forms are transformed into finite elements with those obtained by linear or curve fitting techniques, it is obvious that there will be a difference between the costs of the actual parts of the problem. No matter how much the number of finite elements is increased, the lengths of the paths converted from the curvilinear forms will always need to add some additional lengths to be equal to analog lengths even though the lengths of the paths converge to the analog lengths. Within the scope of modeling shortest path problems with the proposed approach, it is extremely easy to overcome the differences that may occur in the representation of curved forms with finite elements. Because differently from mechanical problems, cost, not form, is important in shortest path problems. In addition to the main vertexes, by adding additional vertexes, it is possible to combine both the original vertexes with two or more straight line, and to provide the original cost values between original vertexes. The additional vertexes that are meant here are vertexes that are added to the original problem but have no effect on topology. In the approach that can be used as shown in Fig. 10, it does not matter whether the number of additional vertexes (virtual nodes) added to the system or where they are located, it is important that the cost between the two original vertexes is fully achieved.

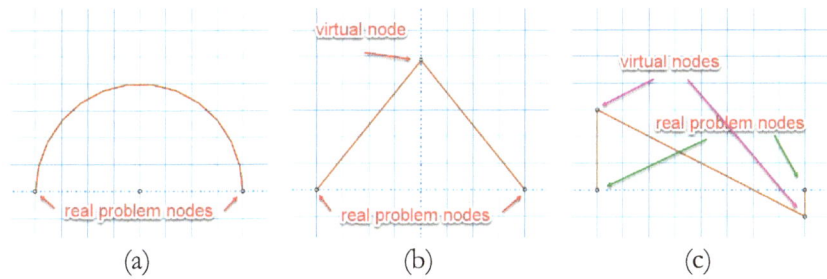

Fig. 10 a Analog form, **b** discretized with single additional virtual node, **c** discretized with multiple additional virtual nodes

With this approach, there are unlimited options in which paths connecting vertexes can be discretized with linear elements without loss.

3.5 Vertex locations and relevance of cost

When transforming the graph of a shortest path problem into a scaled structural model, it is almost impossible to determine realistic positions for vertexes if their costs are expressed in scaled linear straight lines. The total length of path between vertices of transformed system should be equal to the related cost of the original shortest path problem. But sometimes the distance between two points could be less than cost itself, in such a situation the distance can be covered by use of curved lines or multiple straight lines. If the warnings about the curvilinear paths in Sect. 3.4 are taken into consideration, providing the necessary cost with more than one linear straight line should be considered as the most convenient way [8].

4 Shortest path solution of the 3D maze

The geometry of the problem used in structural analyses is presented in Fig. 11, it is aimed to obtain the shortest path among vertices 1 and 9. Although the system also has a 3-dimensional structure, it can be easily examined with the approach previously used in 2-dimensional systems. The system is composed of 9 vertices and 16 lines. In this section, it is intended to make a different sample from the conventional graphical representations shown in planar form. Such different geometric forms may be found in problems with real geographical locations, but the main reason for using this form in the example is to demonstrate the solution ability of the proposed approach.

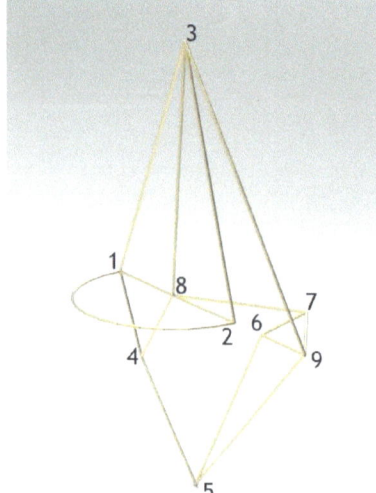

Vertex coordinates

Node number	x	y	z
1	-100	0	0
2	0	0	0
3	-40	120	0
4	-80	-40	0
5	-30	-90	0
6	20	0	0
7	30	10	-17.320508
8	-50	0	0
9	50	0	0

Connection points of sub-paths

1-3	8-3
6-3	9-3
8-7	6-7
9-7	1-8
8-2	6-9
1-2	1-4
8-4	4-5
6-5	9-5

Fig. 11 Scaled visual of the maze belonging to shortest path problem (Units SI-mm)

In order to find the shortest path in the system, pinned support and constant displacement velocity are used at relevant points. The structural model including the boundary conditions and FE mesh is given in Fig. 12.

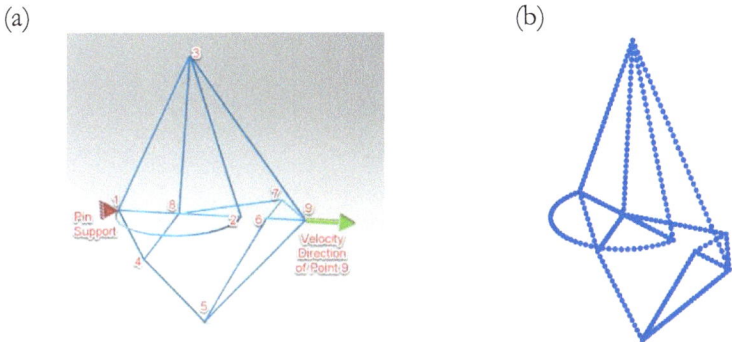

Fig 12 a Boundary conditions of mechanical analogic model **b** FE mesh (approximate element size; 4mm)

The scenario considered for the solution is that during the movement of the point 9 in the specified direction, the elements that will be positioned on the imaginary straight line connecting the points 1 and 9 will form the solution set. In other words, the parameters which should be followed for determination of the set of solution are the relative locations of elements based on displacements.

Table 2 includes information including the maze structure transformed into a structural system by means of a mechanical analogy and parameters used in application of FEM.

Table 2 Boundary conditions, material and mesh definitions. (SI-mm)

Properties of structural model and values of variables	
Velocity of node 9	25
Young Modulus	900000
Poisson's Ratio	0.3
Truss Section Area	2
Number of Nodes	332
Number of Elements	339
Element Type	3D Truss Element
Approximate Element Size	4
Optional Variables and Values	
Damping (Alpha)	20
Density	8e-8

In the analysis where the effect of gravity was not used, the solution of the system was obtained by based on vertex 9 getting away from support node 1 by a velocity of 25mm/sec is analyzed. Appearance variation of system are presented in time domain shown in Fig. 13.

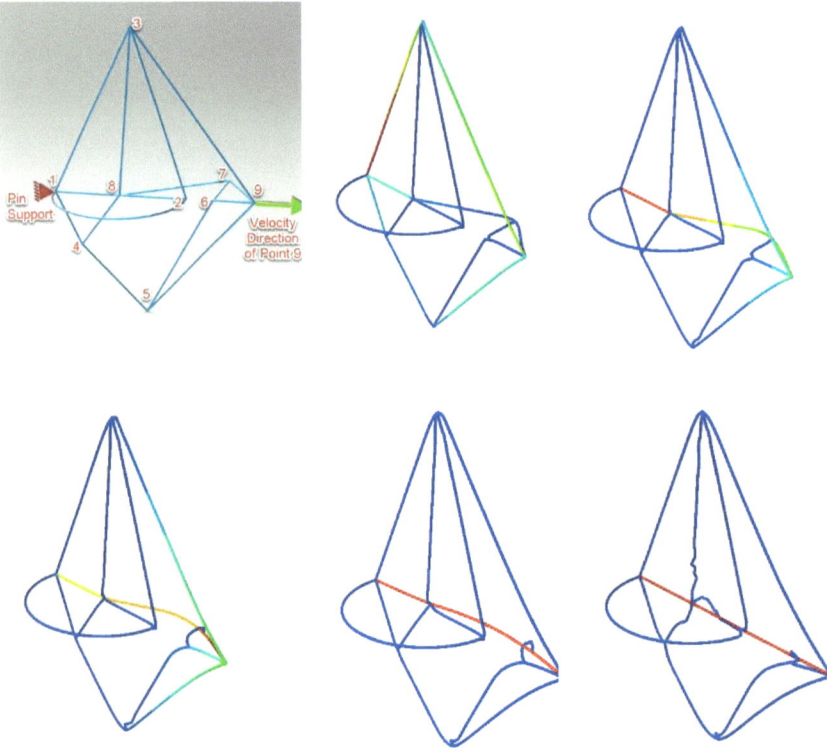

Fig. 13 Variation of appearance of structural system in time domain

When the outputs presented in Fig. 13 are examined, it is seen that the elements of the solution set are located on the line between two boundary condition points. As a result, the solution set consists of the lines 1-8, 8-7 and 7-9 lines in terms of numbering shown in Fig. 12a. Although the positions of the elements were followed in obtaining the solution, it is clear that the elements of the solution set differ from the rest of the system in the stress values.

As mentioned earlier, it should be remembered that stress-based approach may not be reliable in maze solutions due to various factors. In this section, analysis procedure and solution in modeling phase is very promptly obtained since the acceleration of gravity and tensile force are not utilized.

Final stage of system solved including gravity effect is shown in Fig. 14. The set of solution is clearly visible, although it is not pronounced for this problem, in problems of greater systems with gravity effect, excessive deformation values can hinder the determination of the set of solution, as mentioned earlier.

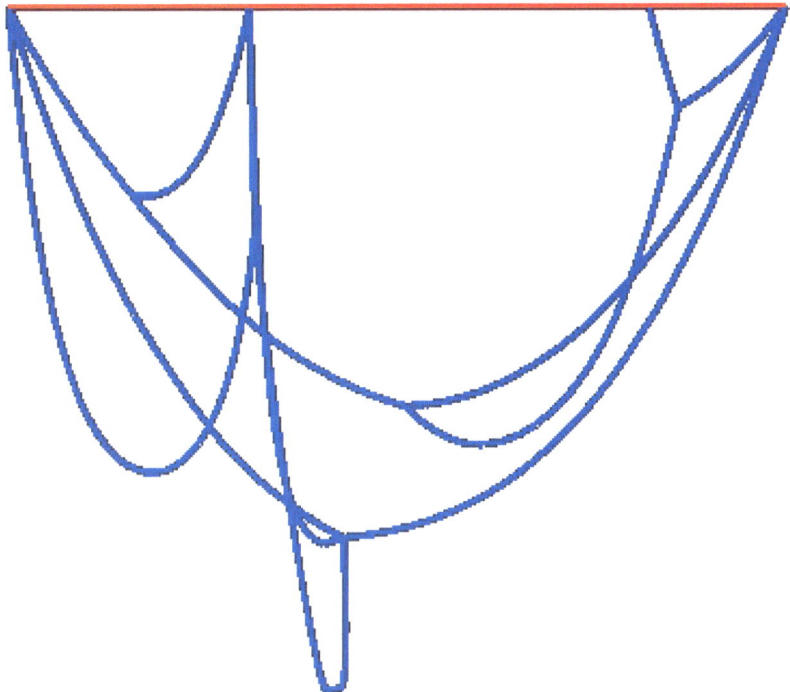

Fig. 14 Solution of the maze system with gravity effects.

5 Comparison of two- and three-dimensional solutions

Thanks to absence of interaction property, since the parts or zones of the body in same location are not in interaction with each other, all the 3D mazes and non-planar mazes can be analyzed as in 2D. Fig. 15a demonstrates 3D maze solved in Sect. 4. Same problem can be solved as in Fig. 15b. All the vertices and paths are shown in same plane. In planar form, the connection hierarchy among path and vertices are same as those in 3D form. Thanks to absence of interaction property, except the connectional interactions through nodes defined in stiffness matrix of FEM, none of the elements are in mechanical interaction with each other.

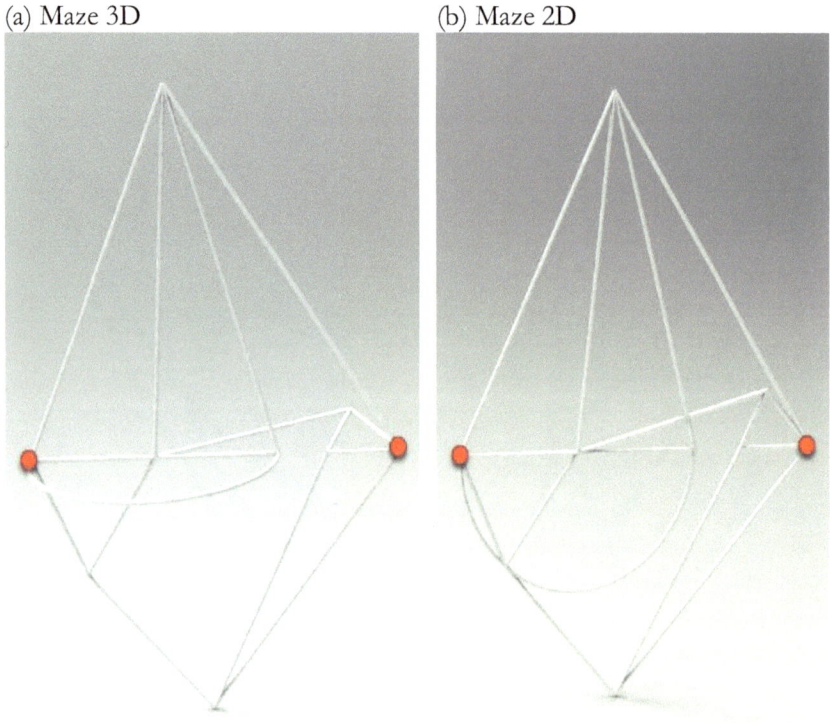

(a) Maze 3D   (b) Maze 2D

Fig. 15 Maze presentations with same line costs and topology

The analyses in this section have the only difference from those in Sect. 4: systems are subjected to gravity effects. The only reason for using gravity is to increase the understandability of the stages that the structural forms of the mazes pass until they reach their final positions. The final forms of two systems with predefined boundary conditions, which are observed after a certain period are given in Fig. 16. Even the initial forms of two systems of which the structural forms are different, subsequent forms belonging to same periods presented in Fig. 16 conclude with the same set of solution.

Exact Solutions of Shortest-Path Problems Based on Mechanical Analogies

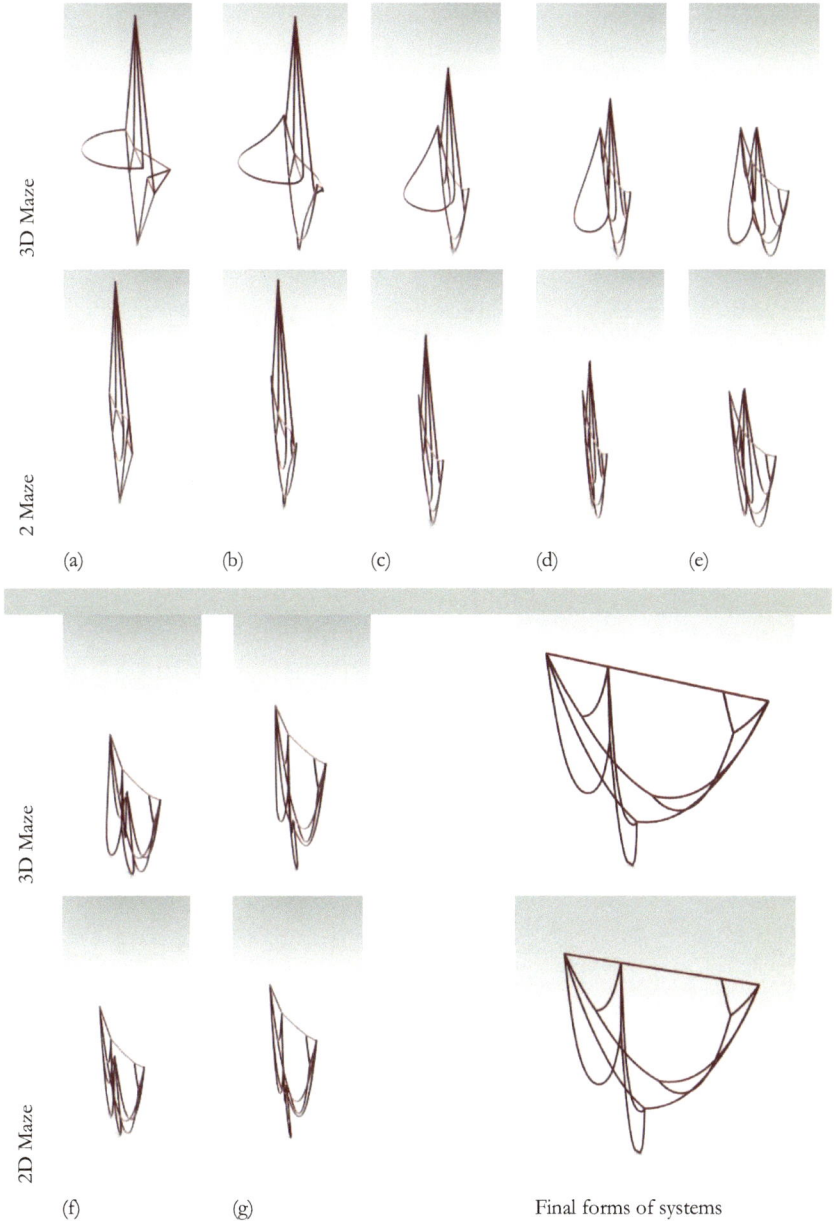

Fig. 16 Solution steps of mazes in time domain

Thanks to inclusion of gravity effect in analysis and neglection of interaction property, not only the set of solutions but also the locations of other paths are obtained to be completely the same. The shortest path solution of two systems are completely the same as the solution obtained in Sect. 4.

6 Shortest path problem solution based on RBD

In this part of the study, it is required to find the shortest paths connecting the vertices 2-10 and 7-8 of the system presented in Fig. 17. The system consists of 10 vertices and 12 lines. Vertexes numbered 1, 4, 5, 8, 9 and 10 are only between two lines, but are intentionally added to follow the behavior. In solution of problem, the motion engine based on RBD rules is used and Unity3D was chosen for simulation environment. The numbered vertexes, as shown in Fig. 17, are connected to each other by the scaled lines with the costs of the real problem. Each path between vertexes of the model created in order to display the physical behavior in a dynamic context consists of appropriate number of rigid trusses each with the length of 10m, which are connected each other by freely rotatable connections. Weight of each rigid truss piece is 10 kg. Since the bodies are not elastic, mechanical properties of materials are not necessary. But a damping parameter used in analysis for aiming to damp the vibrations*. The handling of the damping parameter is entirely in software control of the game engine, and it would be better not to think that it has always been fully scientifically evaluated in the context of RBD rules.

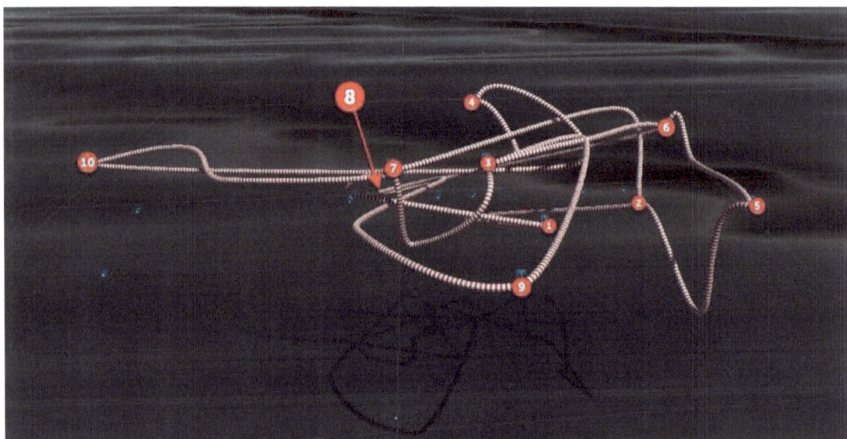

Fig. 17 Randomly located vertexes and scaled lines of maze system

---

* Gravity value and damping damping as percentage of velocity lost per second are chosen as 0.4 m/sn² and 25, respectively.

The connection details and the cost values belonging the lines are tabulated in Table 3. The most of the suggestions and warnings for discretization of FE mesh of examples above are also valid for the model discretization for the solutions of RBD based simulations. As an example, the rules related with minimum number of finite elements which hinder the formation of rigid triangular structures mentioned in Sect. 3.1, is also valid for the RBD based simulations. Apart from calculations by FEM, the gravity can be used safely due to the absence of parameters such as stress and deformation. Discretizing the lines into excessive number of elements and selection of gravity different from zero is only for better comprehensibility and aesthetic concerns in visual presentations.

Table 3 Maze Connection details and costs

Node number	Connected line numbers			
1	7	1		
2	7	9	8	
3	3	2	5	4
4	10	3		
5	6	8		
6	13	6	5	
7	9	1	11	4
8	12	13		
9	2	12		
10	11	10		

Lines	Costs
Line 1	190
Line 2	257
Line 3	108
Line 4	170
Line 5	176
Line 6	177
Line 7	320
Line 8	205
Line 9	305
Line 10	520
Line 11	318
Line 12	424
Line 13	330

The approach used in the previous sections was applied to find the shortest paths connecting the 7-8 and 2-10 vertexes of the system presented in Fig. 18, and the simulations were carried out using Unity3D [9]. In this context, one of the vertexes was turned into a support while the other was subjected to a displacement change in the direction away from the support. Immediately after the solution became apparent, simulation was stopped and the solution set was obtained as in Fig. 18.

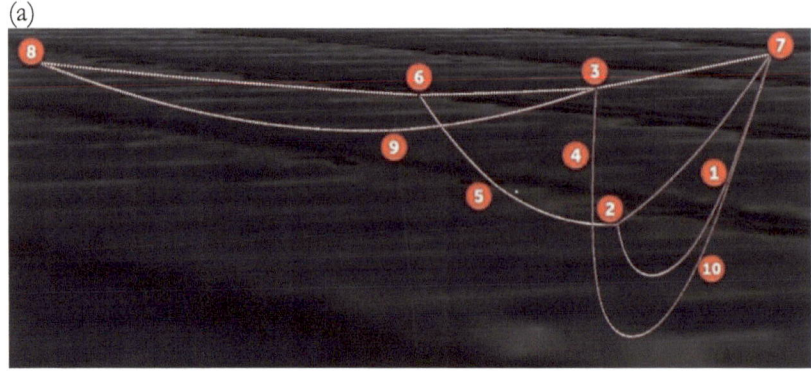
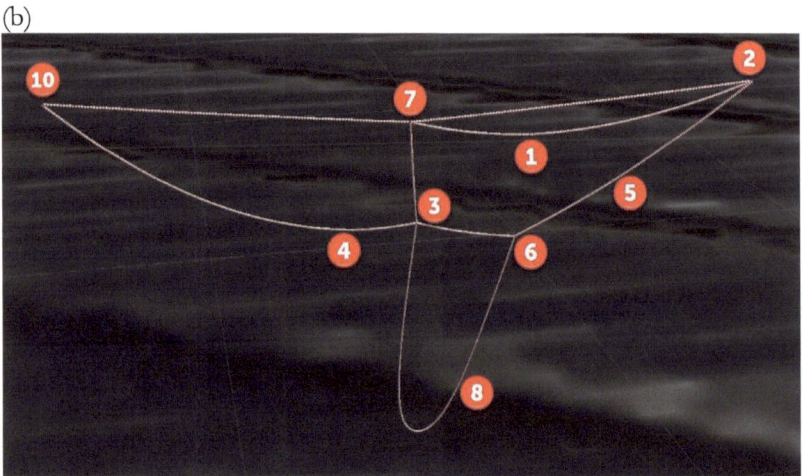

Fig. 18 Solutions of shortest path problems **a** 7-8, **b** 2-10

The RBD based structural model which the maze in original problem was transformed had an arbitrary layout in three-dimensional space. However, the final state of nodes is evolved into a planar projection due to the motion of the nodes and solutions are determined as the top paths of the system. As a result, shortest path connecting nodes 7-8 and 2-10 are obtained as (7-3-6-8) and (2-7-10), respectively. It should be remembered that paths connecting the points are not subject to deformation and interaction is not allowed.

In the scope of solution of shortest path problems, maze systems can be modeled as connected rigid bodies or particles. Behavior of the models based on RBD does not permit modeling of behaviors of theoretically elastic bodies including elongation, stress accumulation, tear, etc. However, based on the forces on joints, these values can artificially be obtained by use of several assumptions. However, accuracy of the outcomes obtained by assuming artificial values may hardly approximate to that of FEM.

The displacements in systems composed of bodies moving in convenience with RBD occur in accordance with geometric constraints. For this reason, similar to systems solved by a simulation based on mechanical analogy, problems investigated in the framework of RBD rules can in many cases be reduced into solution of a geometry problem. At last RBD based models can be used in solutions of planar and also nonplanar graphs.

7 Conclusions

In this study, mechanical analogy-based approaches for the exact solutions of the shortest path problems and several issues to be considered in applications of aforementioned approaches are presented. The presented approaches can be applied with the relevant programs already available or users can easily code their own scripts in this context. To determine the solution sets of examples, approach based on variation of the structural forms was utilized. Mechanical simulations for the follow-up of structural forms, FEM solvers was used for the models consisting of flexible objects and motion engines operating under RBD rules was used for the models consisting of connected rigid bodies.

Due to the fact that it has more details and parameters, FEM is mainly included in the analysis in order to better understand the issue. It should be noted that, although FE Analyses on flexible bodies needs far more parameters, they are potentially applicable to many types of problems. Although not included in the examples of this study, for example, directed graphs can be easily created by very simple assumptions like using the materials with resistant to one-way directed stresses. Similarly, when necessary, structures like logic gates can be constituted by using material properties, interaction or mechanism techniques. Although applications of these approaches were made by dimensional quantities, approaches also have a convenient structure to be coded and specialized only for the solutions of shortest path problems by use of dimensionless quantities.

When the results obtained in the study are analyzed, it is obvious that the mechanical analogies contain the approaches that enable to obtain easily understandable and applicable for the exact solutions of the shortest path problems.

The diversity of approaches and methods available in mechanical analyses provide a wide spectrum of options to be used in solutions, thus, many alternative approaches can be put forward regarding to a specific aim.

References

[1] Biggs NL, Lloyd EK, Wilson RJ (1976) Graph Theory: 1736–1936. Clarendon Press, Oxford
[2] Durnová H (2004) A history of discrete optimization. Mathematics throughout the ages II 51–184
[3] Altintas G (2020) Path Finding in a Labyrinth Based on Displacement of Mechanical Analogical Model. In: Exact Solutions of Shortest-Path Problems Based on Mechanical Analogies. Amazon, pp 29-39
[4] Kocay W, Kreher DL (2005) Graphs, algorithms, and optimization. Chapman & Hall/CRC, Boca Raton
[5] Foulds LR (1992) Graph theory applications, Springer-Verlag, New York
[6] Parker RG, Rardin RL (1988) Discrete optimization. Academic Press, Boston
[7] Gould R (2012) Graph theory. Dover Publications, Inc, Mineola, New York
[8] Altintas G (2020) An Exact Solution Method of The Shortest Path Problems Based on Mechanical Analogy. In: Exact Solutions of Shortest-Path Problems Based on Mechanical Analogies. Amazon, pp 72-85
[9] (2020) Unity (game engine). Wikipedia. https://en.wikipedia.org/wiki/Unity_(game_engine)
[10] Bldwin K (1986) Modern Methods for Automating Finite Element Mesh Generation. American Society of Civil Engineers
[11] Topping BHV (ed) (2004) Finite element mesh generation. Saxe-Coburg Publ, Kippen
[12] Mackerle J (2001) 2D and 3D finite element meshing and remeshing: A bibliography (1990-2001). Engineering Computations 18:1108–1197. https://doi.org/10.1108/EUM0000000006495

[13] Wendt J (ed) (1995) Automatic finite element mesh generation for industrial use. Technical Research Centre of Finland, Espoo

[14] Hamann B, Sarraga RF (1995) Special issue on grid generation, finite elements and geometric design. Computer aided geometric design 12

[15] Ho-Le K (1988) Finite element mesh generation methods: a review and classification. Computer-aided design 20:27–38

[16] Lo D (2015) Finite element mesh generation. CRC Press, Taylor & Francis Group, Boca Raton

An Exact Solution Method of The Shortest Path Problems Based on Mechanical Analogy

Abstract

In this study, an approach based on mechanical analogy is presented for the exact solution of the shortest path problems and the use of many mechanical analysis parameters, which can be difficult to manage, is bound by certain rules. In this manner, by preserving the connection hierarchy and costs of the original problem, it is transformed into a simplified structural form and solution is obtained by use of proposed mechanical analogy. Application of the method by use of Finite Element Method (FEM) or Rigid body dynamics (RBD) is quite easy. In this study, the approach applied by using FEM was developed in such a way that it can be applied to almost all of the shortest path problems of interest in the same systematic process. The presented approach is very suitable for users to write their own codes or the approach can be realized by use of RBD based movement engines or any FEM solver.

The present approach could be considered to be an important alternative for the exact solutions of shortest path problems due to its low computational cost in comparison with the sizes of problems as well as its ease in application and adaptation.

1 Introduction

The algorithms used for shortest path problems can be used not only in problems where the problem-algorithm relationship is easily understandable, such as Road networks, logistic operations and computer networks, but also in the solution of many different types of problems in which there is a parameter to be optimized. Due to the importance of the parameters that need to be optimized in real-life problems, alternatives other than exact solution methods that can also be used in their solutions are undesirable options. Although algorithms used in the exact solutions of shortest path problems are well-known to its details, attaining the solutions may not always be feasible. Because, increasing problem complexity and size comes along with increased computational cost and time to reach a solution. Enormous benefits of the exact solutions in real life problems come along with alternative methods providing the exact solutions, which is still important a matter of research in literature [1-3]. Although their origins go far in the past, use of methods including Depth-First Search, Breadth-First Search, Backtracking, Branch & Bound and A* are still common and research on these methods and their variants are also widely in progress [4-17]. In addition to the approaches based directly on mathematical origins, there are many methods created inspired by biological processes. For instance, methods including evolutionary and swarm optimization algorithms, which were later developed in many additional approaches, have an important place in the solution of optimization problems today [18-24]. Similar to methods constituted by use of analogical approaches including evolutionary and swarm optimization algorithms, a vast amount of methods including algorithms of a variety of processes and phenomena (e.g. biological, physical, chemical or physico-chemical) are still popular research topics of current literature [25, 26, 27]. In addition to the analogical algorithms [28-31] based on the relatively easy-to-understand relationships in the early examples of the methods in this scope, many current algorithms include much more complex analogical relationships. And these algorithms can be used for the solution of very challenging problem types and they still continue to develop rapidly [32-40].

In this study a simple, powerful and fast method based on mechanical analogy is presented to provide the exact solutions to shortest path problems. The approach constituted by taking the warnings in Altintas [41] can be applied for solution of systems composed of connected flexible or rigid bodies. There are no barriers of the use of this approach in solving systems consisting of the combination of the systems mentioned above.

The use of many optional elements, such as the initial location of vertices that are difficult to manage, mentioned in Altintas [41], has been linked by simple rules, and the complexity that can be encountered during the application of the method to large problems has been adopted.

2 Method

In this approach, based on determination of the shortest path between two points of a graph which can be transformed into a scaled structure in a maze form by use of mechanical analogies, the warnings and suggested procedures for application of the analysis are presented in Altintas [41] are still valid. In this context, one of the two points where the shortest path between them is desired is converted into a fixed support and the other is subjected to a displacement at a constant speed in the direction away from this support. Detection of the solution set of the system is determined based on the displacements and the locations of the parts in connection. In this regard, although used approach seems to be the same as the study done by Altintas [41], considering the details and steps presented below, differences of two methods will be clearly understood.

The biggest difference of this approach from that of Altintas [41] is the reorganization of vertices and lines by a simple technique. During this organization, although the topology of the system remains unchanged, all the vertices and lines should be renumbered and relocated due to the new vertexes and lines to enter the system. This approach is thoroughly simple and simplifies the establishment of the system transformed from graph to scaled maze, later from scaled maze into truss system.

By keeping the topology of the system, the approach consisting of transformation of line costs into lengths of geometrical forms is mainly constituted of following steps:

- All the lines are divided from into two equal parts. (This process results in two lines with new lengths and an additional vertex.)
- Vertexes[*] and lines are renumbered.
- The original vertexes in problem are transferred to same locations[†].
- The abscissas of the additional points derived later are positioned as zero, and the ordinates are positioned as half of the cost values of the original line on which the points are located.[‡]

---

[*] In this approach, the term node of finite elements and vertex of graphs can be used interchangeably.

[†] In this study, the point that is accepted as origin in terms of both comprehensibility and organization is accepted as the point where the original vertexes are located.

[‡] Ordinates can be assumed as positive or negative. Problem solution will not be influenced from varying orientation in accordance with geometric limitations, however, the approach is based on selections which are

- Without making any changes in definitions of nodes and lines in its last state, problem is transferred to FE model*. During this process, material properties, cross sectional properties and other necessary values are defined. One of the two points where the shortest path between them should be determined is the support and the other is defined as the loading point†.
- FE analysis is performed without using the interaction feature with the given boundary conditions. Considering the positions of the structural elements during the simulation or in the final state, the solution of the shortest path problem is obtained.

The above items are given in the most general way and can be detailed as desired.

3 Analysis and results

The graph of problem in Fig. 1, it is required to determine the shortest path connecting nodes 3 and 6. The system consists of 8 vertices and 12 lines.

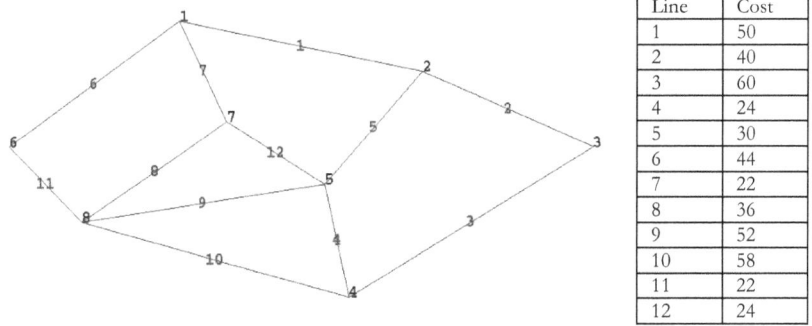

Line	Cost
1	50
2	40
3	60
4	24
5	30
6	44
7	22
8	36
9	52
10	58
11	22
12	24

Fig. 1 Graph representation of problem

---

algorithmically simple and understandable. With this positioning system, original locations will be at the same node, and the additional nodes will be aligned on a line, thereby, there is no need for various coordinate transformations that can complicate the work.

* FEM solution is not compulsory, solution can also be made by RBD based engines.

† These two nodes are at same location with the original nodes of problem. However, as long as the interaction property is not utilized in FEM definitions, truss elements do not interact with each other.

In order to apply the approach described in the Method section to the problem presented in Fig. 1, firstly, all lines are divided from the middle point to two, and the re-numbered vertexes and lines are presented in Fig. 2. As a result of application of the procedure, number of vertices and lines are increased to 20 and 24, respectively.

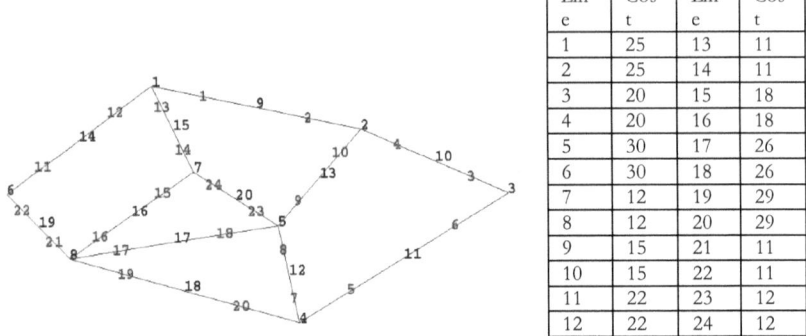

Line	Cost	Line	Cost
1	25	13	11
2	25	14	11
3	20	15	18
4	20	16	18
5	30	17	26
6	30	18	26
7	12	19	29
8	12	20	29
9	15	21	11
10	15	22	11
11	22	23	12
12	22	24	12

Fig. 2 Modified graph of original problem

In order to apply the approach, a FE solution of a scaled model should be performed. The unscaled representation of graph in Fig. 2 can now be scaled easily. Thanks to additional nodes, using bars of lengths proportional to line cots, scaled and planar representations are now possible.

However, it is important to reposition the points, as suggested in the method section, for scaled geometry not to be randomly positioned and for systematic applicability[*]. Applying the given approach, unscaled graph in Fig. 2 is evolved into a scaled FE model constituted of trusses on the same vertical line shown in Fig. 3.

---

[*] Although positioning of vertices by the approach proposed in method section is not the only way to position the model geometry, it is aimed to hinder formation of unlimited number of probable geometries with undesirable special cases. Additionally, there is always a possibility of attaining different systematic solutions.

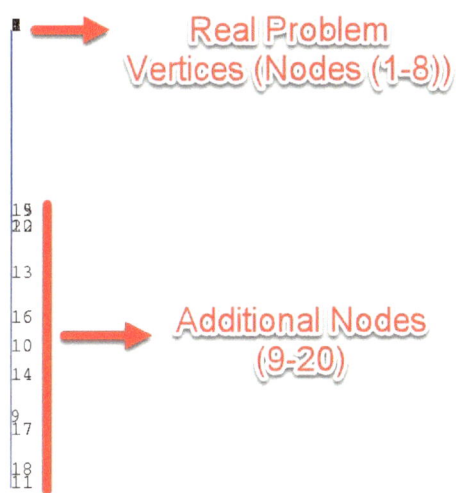

Fig. 3 FE model based on modified graph and coordinates of nodes

The nodes from 1 to 8 in the table in Fig. 3 correspond to as is numbered vertices in original graph representation. Ordinates of remaining nodes are half of the cost values of original lines. The ordinate values of these additional nodes are given with a negative sign, and it should be considered that the positioning of the nodes in this way is not an obligation, but a convenience. This enables a simple arrangement of all nodes and lines on a same line. This approach, which is easy and applicable to coding, makes it possible to degrade the graphs in problem into a geometrical form ready for FE analysis, regardless of the complexity of the shortest path problem under consideration.

A brief look at the input file structure of Abaqus© FE analysis software [42] in Fig. 4 will be useful in terms of model geometry and connection structure as an example for entering FE analysis programs. Information about nodes are connection structures are given in the initial section of file.

The first eight nodes are original vertices of the problem, which are located at origin. Remaining 12 nodes are located at negative ordinate values which are the half of the original line costs.

As long as topology of the system and cost values are preserved, these additional nodes can be arranged in any direction. The second

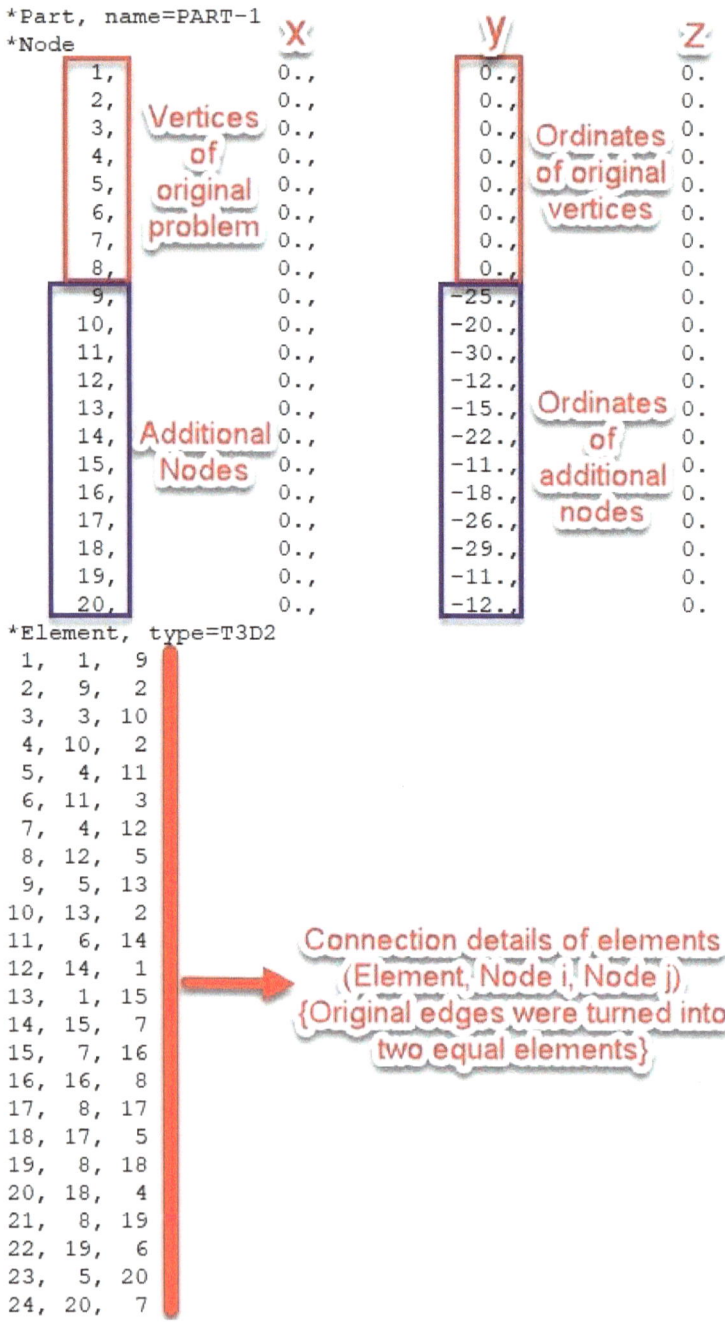

Fig. 4 node and line connection details in Abaqus© input file

part of the file is the definitions of the connection details of lines derived by dividing original lines.

A typical FE input file contains geometric and connection details of the system as in Fig. 4, as well as other necessary information such as material, cross-section and boundary conditions, although the input format is not presented here. Information regarding to material, cross-sections, and boundary conditions are tabulated in Table 1.

Table 1. Material, cross-section, boundary condition and other definitions. (Units SI-mm)

Velocity of node 3	20	Damping (Alpha)	20
Pinned Support Node	6	Density	8e-9
Young's Modulus	200000	Cross Section Area	2
Poisson's Ratio	0.3		

The boundary conditions used in the system are similar to those of Altintas [41], the first of the nodes where the shortest path between them is desired is supported (node 6, pinned support), and the other (node 3) is subjected to a fixed speed displacement. For the stability of the behavior, the roller support is added to the system, which partially restricts the movements of the point subjected to displacement, other than the main movement direction.

The results of the dynamic explicit FE analysis of graph in Fig. 2 are presented in Fig. 5 for given definitions above.

Exact Solutions of Shortest-Path Problems Based on Mechanical Analogies

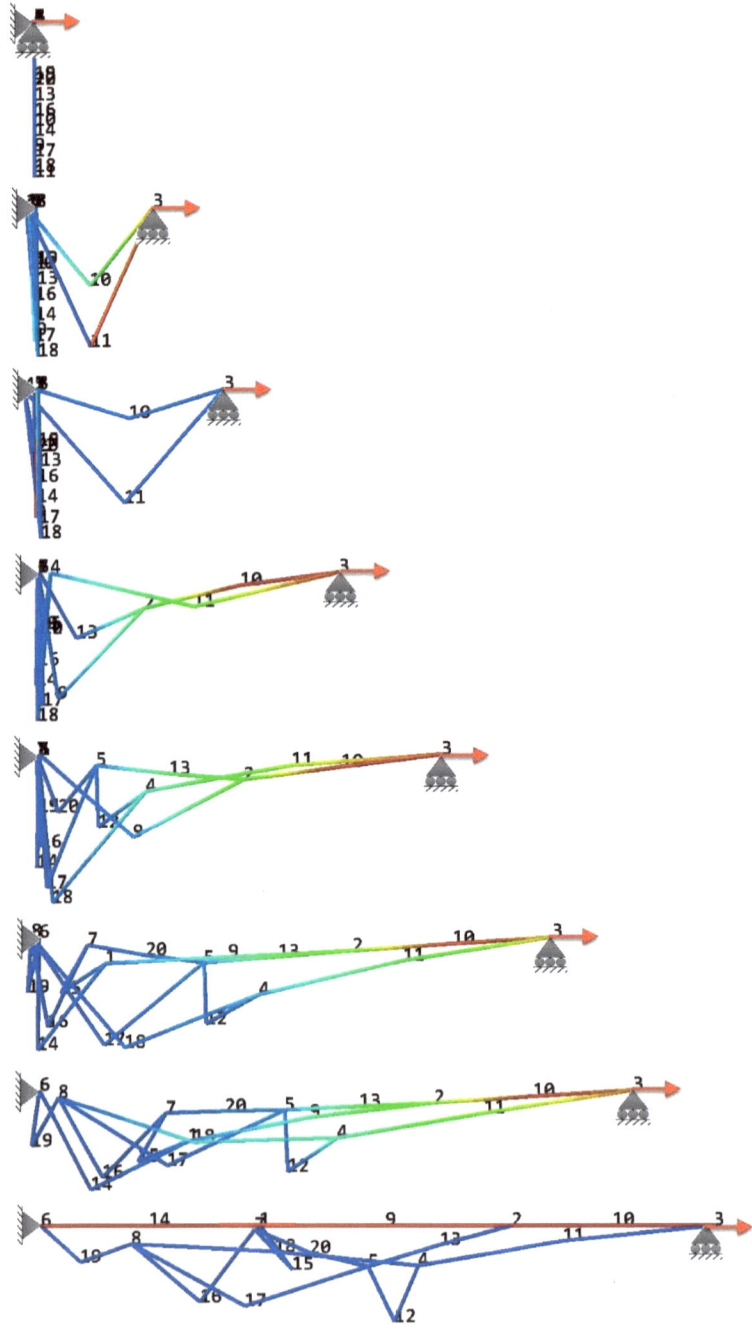

Fig. 5 Variation of appearance of structural system in time domain

For the solution of shortest path problem, shortest path between two nodes of maze formed structure can be seen in bottom image of Fig. 5. The analysis was terminated as soon as the solution set positioned directly on the virtual line between points 3 and 6 where the shortest path between them was desired was cleared. Truss elements constituting the solution set connecting two nodes are determined as 11,12,1,2,4 and 3, based on the numeration of transformed system. According to the numbering of original shortest path graph, solution set can also be expressed as 6, 2 and 1.

4 Conclusions

In this study, a systematic, algorithmic and improvable approach which can be used in the exact solutions of shortest path problems are presented. Owing to this approach, many possible errors mentioned in Altintas [41] have been prevented. FEM was used in analyses and element type in mechanical analogical model is selected as truss elements since it enables to attain a solution using a quite low computational cost. With this approach, analysis can be carried out with almost any FEM software or easily with scripts that can be written for optimization purposes. The familiar physical nature of the mechanical approach used in the analysis enables easy tracking of mechanical simulations and solution steps. There are many parameters that can be included in the solution routine of FEM used as numerical method in the study. The large number of parameters allow analogies to respond to different types of problems if required. For example, although not presented in this study, analogies of directed graph problems can easily be established with parameters that permit one-way strength behavior of the material, mechanism techniques or interaction. Although it does not have the flexibility of the FEM and the ability to respond to different scenarios as a result, RBD based motion engines can be used to apply the presented approach to the shortest path problems within the scope of the study. Because, availability of solution set based on displacement and positions instead of stresses and deformations of overall system removes the dependency on use of FEM in solutions.

Since the main evaluation criterion of the proposed approach in this section is displacements, and in RBD-based simulations, magnitudes such as stress and deformation are not included in the calculation routines, there are even some advantages over FEM analysis. Owing to the fact that it is not used in the interaction property within the approach used, even problems with non-planar graphs or real 3D positions can be easily solved with simple motion engines such as Box2D or FEM using the simplest elements.

Presented approach can efficiently be applied by use of current software or scripts coded by those concerned. The exact solutions of shortest path problems are of utmost importance due to their application areas in real life, and it is considered that the presented approach is a remarkable alternative to currently available methods.

References

[1] Duque D, Lozano L, Medaglia AL (2015) An exact method for the biobjective shortest path problem for large-scale road networks. European Journal of Operational Research 242:788–797

[2] Khani A, Boyles SD (2015) An exact algorithm for the mean–standard deviation shortest path problem. Transportation Research Part B: Methodological 81:252–266

[3] Liu L, Yang J, Mu H, et al (2014) Exact algorithms for multi-criteria multi-modal shortest path with transfer delaying and arriving time-window in urban transit network. Applied Mathematical Modelling 38:2613–2629

[4] Land AH, Doig AG (1960) An Automatic Method of Solving Discrete Programming Problems. Econometrica: Journal of the Econometric Society 497–520

[5] Nilsson N (1965) Some growth and ramification properties of certain integrals on algebraic manifolds. Ark Mat 5:463–476. https://doi.org/10.1007/BF02591142

[6] Green CC, Raphael B (1967) Research on intelligent question-answering system. Stanford Research Inst Menlo Park CA

[7] Hart PE, Nilsson NJ, Raphael B (1968) A formal basis for the heuristic determination of minimum cost paths. IEEE transactions on Systems Science and Cybernetics 4:100–107

[8] Tarjan R (1971) Depth-first search and linear graph algorithms. In: Proceedings of the 12th Annual Symposium on Switching and Automata Theory (swat 1971). IEEE Computer Society, USA, pp 114–121

[9] Ricca F, Faber W, Leone N (2006) A backjumping technique for

disjunctive logic programming. AI Communications 19:155–172

[10] Vukojevic B, Goel N, Kalaichevan K, et al (2007) Power-aware depth first search based georouting in ad hoc and sensor wireless networks. In: 2007 9th IFIP International Conference on Mobile Wireless Communications Networks. IEEE, pp 141–145

[11] Galburt EA, Grill SW, Wiedmann A, et al (2007) Backtracking determines the force sensitivity of RNAP II in a factor-dependent manner. Nature 446:820–823

[12] D'ariano A, Pacciarelli D, Pranzo M (2007) A branch and bound algorithm for scheduling trains in a railway network. European Journal of Operational Research 183:643–657

[13] Lampert CH, Blaschko MB, Hofmann T (2009) Efficient subwindow search: A branch and bound framework for object localization. IEEE transactions on pattern analysis and machine intelligence 31:2129–2142

[14] Utsumi Y, Matsumoto Y, Iwai Y (2009) An efficient branch and bound method for face recognition. In: 2009 IEEE International Conference on Signal and Image Processing Applications. IEEE, pp 156–161

[15] Tao H, Yang Z, Zhang M, et al (2010) A depth-first search algorithm based implementation approach of spanning tree in power system. Power System Technology 34:120–124

[16] Chen X, van Beek P (2011) Conflict-Directed Backjumping Revisited. arXiv preprint arXiv:11060254

[17] Liu X, Gong D (2011) A comparative study of A-star algorithms for search and rescue in perfect maze. In: 2011 International Conference on Electric Information and Control Engineering. IEEE, pp 24–27

[18] Michalewicz Z, Dasgupta D, Le Riche RG, Schoenauer M (1996) Evolutionary algorithms for constrained engineering problems. Computers & Industrial Engineering 30:851–870

[19] Oduguwa V, Tiwari A, Roy R (2005) Evolutionary computing in manufacturing industry: an overview of recent applications. Applied Soft Computing 5:281–299. https://doi.org/10.1016/j.asoc.2004.08.003

[20] Zhang J, Zhan Z, Lin Y, et al (2011) Evolutionary Computation Meets Machine Learning: A Survey. IEEE Comput Intell Mag 6:68–75. https://doi.org/10.1109/MCI.2011.942584

[21] Collange G, Delattre N, Hansen N, et al (2010) Multidisciplinary

optimization in the design of future space launchers, Multidisciplinary design optimization in computational mechanics. pp 459-468

[22] Mora AM, Squillero G (2015) Applications of Evolutionary Computation: 18th European Conference, EvoApplications 2015, Copenhagen, Denmark, April 8-10, 2015, Proceedings. Springer

[23] Wang Z, Sobey A (2020) A comparative review between Genetic Algorithm use in composite optimisation and the state-of-the-art in evolutionary computation. Composite Structures 233:111739. https://doi.org/10.1016/j.compstruct.2019.111739

[24] Sibalija TV (2019) Particle swarm optimisation in designing parameters of manufacturing processes: A review (2008–2018). Applied Soft Computing 84:105743. https://doi.org/10.1016/j.asoc.2019.105743

[25] Adamatzky A (ed) (2016) Advances in Physarum Machines. Springer International Publishing, Cham

[26] Adamatzky A (ed) (2017) Advances in Unconventional Computing-I. Springer International Publishing, Cham

[27] Adamatzky A (ed) (2017) Advances in Unconventional Computing-II. Springer International Publishing, Cham

[28] Hecht H (1939) Schaltschemata und Differentialgleichungen elektrischer und mechanischer Schwingungsgebilde. Leipzig

[29] Gehlshoj B (1947) Electromechanical and electroacoustical analogies. Thesis Copenhagen

[30] Schönfeld JC (1954) Analogy of hydraulic, mechanical, acoustic and electric systems. Appl. Sci. Res. 3:417–450. https://doi.org/10.1007/BF02919918

[31] Bosworth RCL (1948) Thermal mutual inductance. Nature 161:166–167

[32] Endo Y, Ngan CL, Nandiyanto AB, et al (2009) Analysis of fluid permeation through a particle-packed layer using an electric resistance network as an analogy. Powder technology 191:39–46

[33] Chochowski A, Obstawski P (2017) The use of thermal-electric analogy in solar collector thermal state analysis. Renewable and Sustainable Energy Reviews 68:397–409

[34] Goudarzi B, Mohammadmoradi P, Kantzas A (2018) Direct pore-level examination of hydraulic-electric analogy in unconsolidated porous media. Journal of Petroleum Science and Engineering 165:811–820

[35] Siriwardana J, Halgamuge SK (2012) Fast shortest path

optimization inspired by shuttle streaming of physarum polycephalum. In: 2012 IEEE congress on evolutionary computation. IEEE, pp 1–8

[36] Miyaji T, Ohnishi I (2008) Physarum can solve the shortest path problem on riemannian surface mathematically rigourously. International Journal of Pure and Applied Mathematics 47:353–369

[37] Zhang X, Liu Q, Hu Y, et al (2013) An Adaptive Amoeba Algorithm for Shortest Path Tree Computation in Dynamic Graphs. arXiv:13110460 [cs]

[38] Zhang X, Zhang Y, Hu Y, et al (2013) An adaptive amoeba algorithm for constrained shortest paths. Expert Systems with Applications 40:7607–7616

[39] Bonifaci V, Mehlhorn K, Varma G (2012) Physarum can compute shortest paths. Journal of theoretical biology 309:121–133

[40] Rambidi NG, Yakovenchuk D (2001) Chemical reaction-diffusion implementation of finding the shortest paths in a labyrinth. Physical Review E 63:026607

[41] Altintas G (2020) Exact Solution of The Shortest Path in a Maze Based on Mechanical Analogies and Considerations. In: Exact Solutions of Shortest-Path Problems Based on Mechanical Analogies. Amazon, pp 40-71

[42] Smith M (2009) ABAQUS/Standard User's Manual, Version 6.9. Dassault Systèmes Simulia Corp, United States

www.ingramcontent.com/pod-product-compliance
Lightning Source LLC
Chambersburg PA
CBHW040318220526
45473CB00009B/2480